Structures of Broadcasting

Structures of Broadcasting

A symposium

Edited by
E G Wedell

Manchester University Press

Published by
Manchester University Press
316–324 Oxford Road
Manchester M13 9NR

SBN 7190 0388 1

Printed by Spottiswoode, Ballantyne & Co. Ltd., London and Colchester

Contents

Contributors

J. H. Arkell, CBE
Director of Administration, BBC

James Bredin
Managing Director, Border Television

Kenneth Brown
Head of Administration, BBC North Region; formerly Station Manager, BBC Radio Durham

Sir Sydney Caine, KCMG
Formerly Director of the London School of Economics, and Deputy Chairman of the Independent Television Authority. Author of *Paying for TV?*

Denis Forman
Joint Managing Director, Granada Television

Sidney Holgate
Master of Grey College, University of Durham. Chairman of the Advisory Council, BBC Radio Durham

Stuart Hood, MBE
Free-lance Producer. Author of *A Survey of Television*

Alasdair Milne
Controller, BBC Scotland

Philip Purser
Author and Television Critic, *The Sunday Telegraph*

B. C. Sendall, CBE
Deputy Director-General, ITA

The Rt. Hon. John Stonehouse, MP
Minister of Posts and Tele-Communications

E. G. Wedell
Professor of Adult Education, University of Manchester. Author of *Broadcasting and Public Policy*

Introduction

Since 1926 when the essential pattern of British broadcasting was invented broadcasting has come a long way. A single system providing one sound radio service has grown into a dual system providing four nation-wide and eight local radio services, as well as three television services with near-national coverage. Technical innovations as well as the evolving social and cultural patterns of British society are creating new possibilities for, and making new demands on, both radio and television.

How are these new responsibilities to be shouldered and the possibilities realised? Broadcasting, as has been said, is too important to be left to the broadcasters. But even the broadcasters, or most of them, have little opportunity to develop *their* ideas about the future in a context that is at all systematic. The structure which Lord Reith built for the BBC had a high degree of internal consistency and presupposed the acceptance by both broadcasters and their audiences of a strong sense of the moral mission of the medium. The decline of consensus in this and other respects has affected the content but not the structure of broadcasting. Both the BBC and Independent Television are living on their inheritance.

If broadcasting is to be related as creatively to the society of the last quarter of the century as it was to that of its middle years, new thinking must be done, not only by *ad hoc* committees of enquiry, but consistently and along a broad front. If the concept of participation is not to remain an empty phrase, producers and the users and critics of broadcasting must be engaged in this as well as the broadcasting controllers.

New thinking does not presuppose the wholesale abandonment of long-established principles or practices; it does presuppose their open-ended examination in the light of changing needs and circumstances. It is one of the curious by-products of the high degree of concentration of control in broadcasting that it appears to have atrophied discussion of broadcasting policies and structures among both producers and their publics. There is nothing approaching a 'learned society' which concerns itself with the broad range of broadcasting questions. *Contrast*, the one journal devoted to the study of broadcasting in both breadth and depth failed to maintain itself and died as soon as its subsidy was withdrawn.

Thus there has been little conceptual analysis of broadcasting and little opportunity for those working in this field to develop a rationale of their work and to study in depth its social and cultural implications. Nor has the study of the institutions responsible for our mass communications as yet found a firm base. Studies in mass communications have concentrated on effects rather than on causes.

The symposium on 'Structures of broadcasting' held in the University of Manchester sought to make a modest beginning in drawing together broadcasters and those concerned with the effect of their work on the community at large. It proceeded on the assumption that, if new thinking is to be done, there must be a context in which those responsible have an opportunity of identifying their objectives and of establishing criteria of judgment. Such a context must make it easy for them to work together as persons with a common concern which derives from, but is not inhibited by, their organisational or other commitment. For this it is sometimes useful to find a value-free base such as a university can provide. The Manchester symposia on other aspects of social policy had demonstrated that it is both possible and valuable for exponents of conflicting views and objectives to explore these together on the basis of a shared commitment to the service of the community.

The Broadcasting symposium brought together senior broadcasters from the BBC, the ITA and the independent television companies together with academics and researchers, critics, moralists and plain 'consumers' of broadcasting for two days in December 1968. It was surprising to note how many participants, even in so small a professional field as broadcasting, either had not met, or had crossed swords only at a distance.

As the twelve chapters in this book show, the discussion covered a wide range of structural aspects. Mr Stonehouse in chapter 1 looks ahead beyond the structures which may be appropriate to the present limited stage of the communications explosion to the much more comprehensive information transmission, storage and retrieval systems which are currently under study by the development engineers of the Post Office. He also touches on some of the more immediate policy questions, such as the hours of television broadcasting, the fourth channel and the future of local radio.

The two papers on broadcasting finance reproduced in chapters two and three are good examples of the exposition of contrasting positions. Mr Arkell's analysis leading to the identification of the licence fee system as the optimal manner of financing BBC broadcasting differs markedly, both in its presuppositions and its conclusions, from Sir Sydney Caine's arguments for a return to first principles and the derivation from these of a much stronger and more direct demand–supply nexus in broadcasting finance.

Chapters 4, 5 and 6 contain the papers on the role which broadcasting can play in the strengthening of local and sub-regional communities. Dr Holgate and Mr Brown discuss the local radio station in Durham from two aspects: that of the chairman of a station's advisory committee and that of the station manager. Although they were written before the decision to make the local stations permanent and before the publication of the report of the Royal Commission on local Government these accounts of the Durham station are of particular interest since the station was the only one of the eight experimental stations to combine coverage of a town with that of substantial rural area. Mr Bredin's description of the work of Border Television shows that the policy of regional devolution pursued by the Independent Television Authority has been proved to make sense even in comparatively small areas such as that straddling the Border Region. In particular it makes the point that it is not only local origination that is important, but also local control.

It is often argued that regional devolution in broadcasting is incompatible with the achievement of high standards of production. The interesting chapter 7 by Mr Milne makes the case for effective devolution of both power and resources to the BBC regions if the Broadcasting Corporation is to be truly British rather than South-East English in cultural complexion. The companion paper on the problems facing the new Yorkshire Independent Television Company contained stimulating material, but Mr Baverstock was not anxious to have it published, so that it is not reproduced in this volume.

The difficulty of harnessing essentially individualistic artistic talent to the limitations of large organisations has been with the theatre and the film industry for many years. The problem is compounded in broadcasting because of the small number of outlets and by the statutory and organisational constraints within which the broadcasters operate. In chapter 8 Mr Forman sets out the way in which one independent television company has sought to minimise the problem. The analysis of the management structure on Granada's production side constitutes a valuable case study of a field about which far too little is known. Mr Hood complements this in chapter

9 from his exceptionally wide experience in administration (as Controller of BBC Television and Controller of Programmes for Rediffusion Television) and in production as a writer, producer and director.

Chapters 10 and 11 contain papers contributed to a session originally entitled 'The Feed-back Problem'. It was intended to examine some of the ways in which the relationship between producers and their audiences is made reciprocal. Mr Sendall deals mainly with audience research, although he also touches on the role of the advisory bodies used by the broadcasting organisations. Mr Purser explores the role of the broadcasting critic and, incidentally, introduces an unexpected element of polarity by his rejection of the usefulness of most forms of audience research.

In the concluding chapter I examine some of the criteria which may usefully be employed in any forward planning of broadcasting structures, and argue the case for regarding broadcasting as an integral part of the complex network of positive social planning in modern society.

Questions of broadcasting policy are not solved at a two-day meeting. The papers collected in this volume show that the range of presuppositions is wide, even among broadcasters working for the same organisation. Much of the material opens up new ground and sets out terms of reference for future exploration. But the quality of the discussion at the symposium suggests that the debate about 'good broadcasting' and how to achieve it is now able to move into a new phase, one less dominated by the great divide between publicly and commercially financed systems. It is plain that as and when protagonists of particular causes are willing to move outside their carefully prepared positions, they frequently find it possible to meet on a middle ground of fact. They are then able to use this as the point of departure for a common quest of the identification of agreed ends and of mutually acceptable means of achieving these.

I would like here to express my thanks to all my co-contributors, who took a good deal of trouble over their papers, in order to make them ready for publication; and to Mrs Helen Wallace and Mr Peter Sanderson of the Department of Extramural Studies who helped with the assembly and sub-editing of the material.

E.G.W.

1

Broadcasting in the 1970's

John Stonehouse

Anyone who says he knows for certain how television is going to develop and what impact it will have on human society during the remaining thirty years of this century is probably a fool. Anyone who claims to understand how television will shape the next century is certainly an idiot. Those of us who can look back thirty years—and I can just about do that—will agree that long-term prophecy is best left to the astrologers. Who would have imagined in 1938, when television was a mere flicker for a privileged few, that thirty years later it would have become essential household furniture for 90 per cent of our population. Who would have anticipated television broadcasting on a wide scale, not only in the industrial states but in countries like Uganda, Mauritius and Siam? Who would have predicted that hundreds of millions of viewers all over the world would see the Olympic Games from Mexico simultaneously, and that many of them would receive the pictures in perfect colour and perfect clarity.

Of course it is possible—and indeed desirable—to speculate on what will be technically possible by the year 2000 and to prepare our broadcasting authorities for change, but dogmatism and arrogance on this subject, as on most others, can be a straitjacket and the enemy of dynamic progress.

What is clearly true is that the pace of technological change is now quite breathtaking. Solid state devices, microminiaturisation and computer aided design give the inventors new tools, helping to increase the rate of change by at least a factor of four compared to thirty years ago. Within the next thirty years new electronic techniques may dispense with the tube and the 'box' will pass into history like the penny-farthing bike. TV sets could then be hung on the wall like pictures or recessed. Indeed, we may have television wallpaper providing a psychedelic background to suit the mood, when not actually being 'watched', or a viewer involvement in depth when three or four walls are switched on. This is a horrifying thought. At Expo 67 in Montreal there was a cinema in the round which was a preview of what may become possible in homes with wall TV. More likely in the next few years is that we shall have home television recorders which will record the programmes we are not in to see so they can be played back later, or enable us to hire programmes we want to see from libraries. This will have an enormous effect on the TV set as an educational medium; it will really become a teaching machine with the viewer able to play back a lecture on a subject again and again. In the years ahead when re-training will be of such importance a teaching machine in every home will be of immense signifinance.

It is fascinating to speculate on these possibilities, but I will avoid the temptation and limit myself to television today and the next seven years. The year 1976, in fact, provides a convenient watershed as both the British Broadcasting Corporation Charter and the Independent Television Authority lapse on 31 July in that year. There will certainly be a great deal of debate about how sound and television broadcasting should be organised after 1976. I do not want to enter into that debate now, except to say that the Government of the day will almost certainly want to establish some forward-looking inquiry in 1971 or 1972 which would report in time for Parliament to enact new legislation by July 1975.

It would be a pity if preoccupation with what happens after 1976 prevented us from grasping the opportunities available during the next seven years, although of course decisions should not be made which pre-empt the long term review. The 1975 policy will be vitally important; it will lay down the broadcasting framework at least until the end of the century.

Television is a medium which everyone likes to criticise; it has long ousted the weather as the most common talking point. Although all of us can see room for improvement and some valuable suggestions have been made recently, we should, I think, get the subject into perspective. British television is among the best in the world. I cannot say it is the 'best' because I have not seen them all, but it is certainly better than any I have seen. It is generally recognised as being vastly superior to TV in the USA despite the multitude of channels across the Atlantic. Why is this? It is because of the long tradition of public service by the BBC, not sullied by crude commercial considerations, plus the thrusting competitive element provided by the independent programme companies under the benevolent but strict control of the ITA.

We have in Britain, so far, managed to get the framework right. Within this framework the broadcasters have been able to do a remarkable job. One of the fears fifteen years ago was that our TV sets would be full of American-produced programmes. That has been avoided, mainly by the efforts of our own producers. There are of course the Westerns, the old American films and Rowan and Martin's *Laugh-In* with its dated jokes. But these are little more than fill-in material between the excellent programmes made in Britain. On a typical night recently out of eighteen hour's viewing on three channels, less than two hours' originated in the United States. An additional point of pride is that both the BBC and the independent companies—like ATV—have sold many programmes abroad, earning a valuable export income. *The Forsyte Saga* and *The Saint* are just two examples of successful exports. The balance of programmes, considering the restriction on broadcasting hours, is also about right.

The programmes for the same night, for instance, provided an excellent spread of choice. Out of six hours viewing from 5.30 to 11.30 BBC 1 had nearly two hours of news programmes with the rest of the evening given over to comedy, music, thrillers and a home features programme. ITV had over an hour of news programmes, *Coronation Street*, and one and a half hours of theatre (two and half hours, by the way, networked from Granada). I pay tribute to the influence of the non-metropolitan companies. BBC 2 provided a balanced programme ranging from an art profile to Marty Feldman, all in colour.

I am not suggesting that there is no room for improvement in TV programmes and programming. There could be more co-ordination in timing between the BBC and ITA, for instance in sports, and agreement between them on methods of audience rating as the present two separate techniques are irritating and confusing to the public. But I do believe that those who attack the quality of British television indiscriminately do a grave injustice to our broadcasters who perform a valuable function in our society.

Some of those who criticise what is done with the medium, either do not watch television or assume for it a role which it should not have. It is not concerned primarily with opening out opportunities for democratic participation: its principal role is to entertain; providing information and education are secondary, though sometimes integrated. As Mr Gerald Kaufman has pointed out in a recent edition of *The Listener*, the ethos of British broadcasting is quite different from that in the United States in that the viewer here usually accepts what he sees as impartial. The information providing role is therefore absolutely vital. The manipulation of TV news by vested interests could do grave damage to our democracy. There must be fair opportunities for various views to be ventilated on TV, and interviewers must not rig the programmes merely to obtain exciting and compulsive viewing.

It has been suggested that the medium is so important that a Broadcasting Council similar to the Press Council should be established to supervise it. The differences between the Press and TV media, however, are considerable. There are dozens of national and provincial newspapers but only two TV networks. The Press is completely independent of statutory authorities whereas the two TV networks are controlled by a Corporation and an Authority, both ultimately responsible to Parliament. The BBC Board of Governors are not themselves broadcasters; they employ and supervise broadcasters and are expected to ensure that programmes are balanced. The ITA Board are not themselves broadcasters, but they can regulate the programmes of the broadcasting contractors as to impartiality, taste and balance. There have recently been several examples of the BBC Governors or the ITA Board asserting themselves against the wishes of the broadcasters—control over interviews of children is an example.

To superimpose another statutory body on top of the two that already exist is neither necessary nor desirable. It would lead to bureaucracy and dilution of responsibility and could curb the creative genius which we want from broadcasters. If public opinion feels strongly about the way the authorities are carrying out their responsibilities there are ways in which these views can be expressed—perhaps a useful addition would be a TV magazine programme on which viewers could make proposals about programme content and for new programmes. This could give viewers an opportunity similar to that which they had on the useful *Talk Back* programme promoted by the BBC.

There is certainly room for more experimentation with the television medium and the BBC and the programme companies should be given every encouragement to develop new ideas. One of the obstacles in their way is the limitation on broadcasting hours. The Government is not yet in a position to announce any extension of TV broadcasting hours, as we will want to weigh up the overall extra costs to the economy before adding to the extra hours which I announced in July 1968 and which are now in use. On the other hand, there are powerful arguments not merely to extend hours but to abolish restrictions on TV broadcasting hours altogether. Such a de-restriction would not of course mean that the BBC and ITV would be broadcasting twenty four hours a day. With certain safeguards they could choose what to do with the unlimited hours. It would (1) give the broadcasters far more flexibility than they now enjoy, (2) help them to promote experimental programmes and magazines for minority groups, (3) allow peak-hour programmes to be re-broadcast at off-peak hours for shift workers and (4) allow news programmes at breakfast, by which time a lot of visual material has accumulated, particularly from the Western Hemisphere.

The Government must also consider shortly whether to allow a fourth television channel and when. In the next year or so there will be dramatic developments as colour on UHF comes into operation on BBC 1 and ITV. Colour introduces a completely new dimension to TV and its impact will be very great indeed. The Government will have to consider if it would be in the national interest to introduce the fourth channel within a year of universal colour—that is in about two to three years' time. On the one hand there will be an extra burden on national resources; additional programmes will have to be produced and extra overhead expenditure will be incurred, although part of the capital expense would already be covered by the network of UHF broadcasting centres now under construction. As against that if an additional channel went to commercial television (thus maintaining the balance with BBC, already with two) it could provide an excellent opportunity for existing and new independent companies to experiment even more with adventurous programmes at peak as well as off-peak hours. There would need to be safeguards to ensure that an additional channel did not drag programming down to the lowest common denominator in the competitive search for high ratings. The recent experience with three channels is encouraging: good quality has been maintained despite intense competition.

I have not attempted here to make a comprehensive statement of the Government's broadcasting policy. There are several fields I have not even touched, such as sound broadcasting, for instance, which is very exciting and important and where we await the full review in July 1969 of the BBC's local radio experiment. Perhaps I should add that in the meantime we are likely to turn down all requests for commercial radio, including those from municipal authorities.

2

Broadcasting finance: the licence fee system

J H Arkel

In this discussion of 'Broadcasting Finance', I should like at once to make plain where I stand. I hope it will not be thought that I am making a political statement when I say in regard to this subject that I am a conservative—I like things the way they are. So I am not going to perform that tempting academic exercise of making simple things complicated—of showing how, if broadcasting did not exist, it might most tidily be invented—but to carry out the much less exciting, much less headline-catching, but nevertheless valid function of justifying the *status quo*.

What is the *status quo*? The Prime Minister, in a speech at the opening of Yorkshire Television in the Summer of 1968, described it as involving 'a national television service based on two diametrically opposed philosophies . . . The two television systems', Mr Wilson said, 'are in fact complementary parts of a single public service . . . Each of the two television services has drawn many of the best qualities of the other to itself and each acts as a guardian against abuse or excess by the other.' An essential difference between these two services is of course that one is financed by advertising and the other by the licence system.

May I make three other points clear. First, I shall not be exploring such interesting by-ways as the relative cost of different types of programmes, though that is obviously indirectly relevant to the size of the licence. Second, as I have no doubt we are agreed here, it would not be feasible, even if we thought it desirable, drastically to reduce the overall amount of broadcasting, though if the hours of television were so reduced, we could manage with a smaller licence fee. The trend appears to be in the opposite direction. Third, I do not intend to spend time discussing the machinery by which the BBC controls its expenditure, although its efficiency obviously

2

affects the amount we spend. When in 1968 we called in the American business consultants, McKinsey's, some people suggested that we did so because we were badly in the red or because we were a prey to the inefficiency resulting from an inflated bureaucracy. Neither of these things was true. McKinsey's were a natural follow-on in a sequence of work done over the last ten years by both our own internal management consultants and outside consultants—In the Autumn of 1967, for instance, I brought in Urwick Diebold to advise me on our Computer Structure. The public have a right to know whether we make the best economic use of our resources, and we believe ourselves to be reasonably efficient, a belief endorsed by successive committees of enquiry including Pilkington. And McKinsey's have in fact stated in their report that, in general, the BBC does a good job of planning and controlling the use of available resources. But we are not so complacent as to suppose that there is not always room for improvement, particularly in a large organisation and in the long term.

And so to my main subject, which is: How best to find the money to pay for broadcasting on the present, or an increased, scale? Let me look in turn at some of the alternatives. If I say that I intend to do so in ascending order of acceptability, taking the least acceptable method first, it will not perhaps surprise you that I begin with Pay-Television.

Pay-TV

I will deal with Pay-TV briefly, since the experiment which began in London in January 1966 and in Sheffield in November 1966 has now been wound up. The Pilkington Committee recommended against such an experiment because the system is by its nature exclusive and socially unjust. Indeed it is alien to the whole concept of a public service. For technical reasons Pay-TV can never be more than a minority service, and an urban service for the well-to-do at that. Country-dwellers, people living in communities too small to be worth Pay-TV's while to service, old-age-pensioners and others who find it as much as they can do to afford 'ordinary' television—none of these are ever likely to receive it. Lord Pilkington put it vividly in a recent letter to *The Times*, when he wrote that 'since Pay-TV is basically expensive, it would have to take out the plums in order to attract shilling-in-the-slot viewers and could never provide a fully balanced programme service'. The White Paper on broadcasting of December 1966 stated that 'the government would not allow a situation to arise in which the vast majority of viewers were denied the viewing of major sporting events'. But this exclusivity is of course exactly what would happen if Pay-TV were ever in a position to out-bid the BBC or ITV for the right to cover such events, so that instead of widening the range of programmes available to the ordinary viewer it would actually restrict it.

It is worth recalling here the claims made on behalf of Pay-TV before the Pilkington Committee, claims for instance that it would provide 'programme items of a kind or quality which services financed . . . by licence or by advertising revenue could not, or would not provide'. More recently the consortium involved in the recent experiment claimed to have supplied 'lively sports coverage, up-to-the-minute films and a wide choice of programmes aimed at minority audiences'. An analysis of what has actually been provided hardly bears out these assertions. The main diet has been feature films or sporting events, with one or two exceptions. There have been very few documentaries, or plays or programmes catering for minorities. Of ninety programmes transmitted on Pay-TV, in one four-week period, seventy-four were feature films, most of which were shown not just once or twice but four or five times. Many of these were of a not particularly edifying type and they included some which the BBC had actually rejected. Few were superior to what was freely available on BBC and ITV, but where they did differ was in the cost, six shillings a time, which would have effectively put them beyond the reach of many viewers. 'Pay' is indeed the right word for this type of television service and that related word 'expense'. Pay-TV is provided at the expense of the ordinary viewer in favour of the wealthier household, at the expense of those living in thinly populated areas in favour of those concentrated in the cities, and for the popular taste among its own subscribers at the expense of the minority taste.

Advertising

Let me now turn to a second alternative, advertising. On the general question of financing broadcasting in this way, I cannot do better than quote the opinion of Lord Hill, who has said that 'we have reached a satisfactory position in this country in the matter of broadcasting, in which we have a commercial service competing with a public service'. The BBC does not say that broadcasting should never be financed by advertising; what it does say is that BBC broadcasting should not be financed in this way. The essence of public service broadcasting is that it should serve one master only, *the public*. A service which carries advertisements must inevitably, and indeed rightly, have other obligations too, to the shareholders and the advertisers. These must in some degree affect its output and programming. When commercial television was first contemplated in this country, one of the arguments advanced to justify it was that the BBC would still be there to set high standards. It would be ironic and tragic if the resulting achievements of ITV were now to be used as an argument for forcing advertising upon the BBC. It might be unfortunate too for ITV itself. If the BBC offered an alternative means of advertising, the position of the ITA in maintaining standards, both in programming and in the content of commercials, would undoubtedly be weakened through the battle for advertising revenue. We feel too that we have an obligation to the public, as well as to the writers, performers and producers of our programmes, to ensure that this one area of communication is kept clear from the sometimes strident voice of the advertiser.

Most of these arguments apply with equal or greater force to other types of finance through commercials. We do not for instance believe the public would wish to see in this country the direct sponsoring of programmes on the lines followed in the United States. Nor do we think that a 'mixed economy', with the licence fee kept unnaturally low by means of an advertising subsidy, is desirable in theory, even if Canada had not found it unsatisfactory in practice.

This suggestion in its most persuasive form is that, while television should be left inviolate, some part of the Radio Services should be forced, or permitted, to carry advertising.

Let us be clear what this means. It would not be possible to confine advertising to Radio One, the favoured candidate for it, since Radio One is at present linked for much of the day to Radio Two, so we would have at once two commercial channels not one. The advertising income from these two channels, taking the most optimistic estimate, would not go very far to meet the BBC's expenditure even in Radio. More serious still, we believe that once we had accepted advertising anywhere, it would sooner or later become impossible to prevent it spreading inexorably to the other radio channels and thence to television. It would be the thin end of a very big wedge. Future governments would have an unanswerable argument for refusing to face up to the political inexpediency of agreeing to any further increase in the licence fee. In a phrase, there would be 'a fundamental change of policy by a side wind'.

Government subvention

So much for Pay-TV and advertising. A third possible source of income is direct Government subvention from taxation. Let me first dispose of the irrelevant argument that because this system works reasonably well in the case of the BBC External Services it could also be applied to domestic broadcasting. There are essential differences between the two. In the former the BBC is over the long term carrying out aims on which both the Government and the BBC agree, and the functions of the two are distinct. Equally important, the present system functions well because the External Services operate within the BBC, whose editorial independence is safeguarded by the licence fee. It should be remembered too that the cost of the External Services is relatively small—£10 million a year—compared to the annual budget of the domestic services of nearly £79 million. No Chancellor will readily find that sort of money from taxation or if he did Parliament would want to take a *controlling* interest in how a sum of that size was spent.

The arguments against a subsidy for domestic broadcasting were well put by the Treasury in its evidence to the Pilkington Committee, when it pointed out that, 'If the Corporation were to be financed from general taxation, instead of from licence fees, the Corporation would be in a position very similar to that of a government department and would have to become subject to much the same kind of detailed control . . . Parliament would regard the minister responsible for the Vote as being in some degree at any rate accountable for the Corporation's policies and actions.' In other words, if the BBC were financed through taxation it would cease to be independent, and would lose that freedom which is the essential climate for broadcasting as a public service. Almost as serious, the BBC would never have a guaranteed, long-term source of income, exempt from the whims of political chance, and without such an income no long-term planning is possible—efficient financial long-term planning included.

But to say that we would not welcome the taxpayer's money to finance all our domestic services is not necessarily to rule out for ever the possibility of accepting it for some. Many viewers resent paying for school broadcasting, and argue that it should be financed by those who benefit from it, i.e. the Department of Education, the local authorities, or the schools themselves. We have indeed entered into not dissimilar arrangements with the Open University, where we are providing the expert knowledge, and the staff and facilities to make programmes, while the University is drawing up the syllabus and footing the bill.

Other sources of income

I want now to deal with what I have rather vaguely put down as 'other sources of income'. Our only experience of these arises from local radio. The Government White Paper on broadcasting of December 1966, which authorised a nine-station local radio experiment on VHF, proposed that 'its income should derive so far as possible from local sources and not from a general licence fee'. A general subvention from the rates was ruled out. Other possible benefactors mentioned were 'the local university, where there is one; Chambers of Trade and Commerce; local Councils of Churches; arts associations; and other representative bodies active in the social and cultural life of the community'. It is too early to give a decisive verdict on the extent to which these hopes have materialised, and in any case the Postmaster General will not be reviewing the local radio experiment until next July. So far the BBC has met the capital cost of setting up each station and is meeting where necessary a part of the running cost. Some local authorities have made substantial contributions. A number of companies and organisations have given smaller donations. We feel the arguments against accepting advertising on local radio are just as compelling as against financing the networks in this way. At some stage, it may be, the question of paying for local radio partly or wholly from the licence fee will have to be considered. This would of course raise the further question as to whether it is reasonable to expect people living outside the cities to finance a service they cannot enjoy. As to that, one can only point to the assumption, which is implicit in the present licence fee system, that all will contribute to the enjoyment of each. Certainly these are difficult questions, but we would hope they would be resolved before the future of local radio is considered next year.

The licence system

And so I come to the existing and, as I believe, by far the best method of financing broadcasting as we, the BBC, conceive of it: I mean the licence system. The wireless licence is even older than the BBC, for it was first introduced on 1st November 1922. Thirty-five years later, in 1957, the Treasury, no less, made the following statement in evidence to the Public Accounts Committee:

> The existing system which has been endorsed by successive Committees of Enquiry over the years, is sufficiently flexible to obviate excessive accumulations of reserves by the Corporation over long periods. At the same time, it enables the Corporation to plan ahead and preserves the independence the Corporation has always had in the day-to-day conduct of its business. In the opinion of Their Lordships (i.e. The Treasury) it would be undesirable to break the link between licence revenue and the annual grants to the Corporation.

In 1962 Pilkington reached the same conclusion. In the face of such evidence it ought not to be necessary to justify the licence system. We do not believe that there is widespread hostility to it as a system, though of course everyone grumbles a little when the fee goes up, just like everything else. A few years ago, when we changed the format of a popular five-minute religious programme, *Lift Up Your Hearts*, 4,500 people wrote in to complain. The ending of *Children's Hour* on radio produced 3,500 complaints. The raising of the combined licence to £5, in April 1965, brought in, in that month, fifty letters of complaint with a few more following later. The ending of the serial *Compact* yielded 1,500 complaints. The threatened demise of a favourite Sunday morning radio programme, *As Prescribed*, in our South and West Region led to a petition with 49,000 signatures. Yet by contrast the proposed increase in the licence fee to £6, announced in July 1968, brought in no more than 100 or so letters of complaint.

What of the cost of the licence, at £6 in this country? Compare it with £9 6*s*. o*d*, in Sweden; £9 13*s* o*d* in Austria and £9 15*s* o*d* in Switzerland. And it is the same story throughout Europe, where in many cases they are paying considerably more for fewer hours of broadcasting than is provided here by the BBC alone.

What of the suggestion that the licence represents an excessive and burdensome impost on the average family? To rent the cheapest television set obtainable today costs rather over 8*s* a week; larger ones cost up to 12*s*. The licence, at £6, costs under 2*s* 4*d*. Under Government regulations at least seven months' rental has to be paid in advance, around £14 or more. A good outdoor aerial costs another £10. Who are these mysterious families who can find nearly £25 to install a set, but not £6 to pay for the programmes they receive?

The British are great newspaper readers, and an average family which takes one daily newspaper at 5*d*, and one Sunday newspaper at 7*d*, will spend on these alone £8 6*s* o*d* a year. It hardly makes sense to say they can manage that but not £6 for a radio and television licence.

Still one may hear the argument, 'But the licence is always going up.' Again, let us look at the facts. In 1922 the average industrial gross earnings for men were £2 16*s* 3*d* per week; a radio licence cost 10*s* a year or under 2½*d* a week. Until 1946 the licence remained the same. Then it was raised to £1 or less than 5*d* a week. But by now the average wage was £6 o*s* 9*d*. At the same time, a combined radio and television licence costing £2 a year was introduced. In 1965, after an intermediate rise to £4, the licence rose to £5 or less than 2*s* a week, while wages were now £25 8*s* 11*d* a week. In 1969 the licence fee was £6 but the average earnings for 1967 were £27 18*s* 1*d*. and the January 1969 figure will no doubt be higher still. But throughout this long period of time between 1922 and 1969, the licence fee expressed as a percentage of a man's weekly earnings was always about one third of 1 per cent, or put in another way for every £1 a man earned his licence cost him less than 1*d*. To say therefore that the licence fee is becoming too expensive—if anyone does say that—is in a sense to say that everything is becoming too expensive, because the licence fee does not rise proportionately more than wages; and if you were able to extract from the figures the cost of the extra listening and viewing facilities provided (BBC 1, BBC 2 and four radio channels) then I feel confident that the licence fee would not have risen proportionately as much as the cost of living.

This might be the appropriate point to mention what I might call the Wedellian heresy. I am not referring here to Professor Wedell's proposal to redraw the broadcasting map of the country and to remodel the BBC by forcing it into a structure not unlike that of the commercial programme companies. The basic organisation of broadcasting is outside the scope of this lecture, and I will say of this suggestion only that it has at least the attraction of novelty. What I will comment on is Professor Wedell's suggestion that money raised by a levy on commercial television should go into a common fund and then, via the Treasury, be handed over to public service broadcasting organisations, of which the mutilated remains of the BBC would be part.* Speaking for the BBC, we can, quite frankly, see little to commend these proposals. As Professor Wedell will recall, we pointed out at the time that they would at once introduce two variables, the fluctuating levy from the programme companies and the uncertain inclination of the Treasury to part with the

*It is fair to say however that Professor Wedell emphasised in his book "Broadcasting and Public Policy" the need to devise a means by which the fund could be directed into what might be called a general broadcasting fund, and paid to an independent body established to administer the funds.

money. The result would be of course to place the BBC's finances at the mercy of precisely the pressures I have already described. Dependence on the Government for money would, we believe, lead to dependence on it in other ways. The licence system, offering the BBC an independent source of income, and thus an independent existence free from Government or commercial pressure is, we still firmly believe, superior in every way. Nor do we think that to blur the line between two different types of broadcasting authority, one financed by the licence, one by advertising, and each competing with the other, would necessarily be in the public interest.

The BBC's finances today

Of course, we do our best to keep the licence level down by earning money from commercial activities. Non-broadcasting activities, mainly BBC Publications, brought in a profit of £945,000 last year. BBC Enterprises, which sell discs and films of BBC programmes, is another and, we hope, increasingly profitable activity, with an annual turnover of about £1½ million. This year some of this will be in roubles, from, of all things, the Russians watching *The Forsyte Saga*. But of course such activities can never obviate the need for a much larger and more important source of finance.

Of greater immediate importance has been the Post Office campaign against licence evasion, launched two years ago, which is believed to have cut our annual loss of revenue from about £10 million to about £6¼ million now. Another £6 million would of course be a useful addition to our domestic services budget, which in the last financial year involved around £78 million, of which operating expenditure accounted for £67 million and capital expenditure for £11 million.

Here I must draw attention to an important characteristic of broadcasting finance, that increased capital expenditure does not, as in most businesses, yield a higher return in annual earnings, except in the special case of colour television. The licence fee is the same whether or not the licence-holder can receive perfectly our four radio services and two television channels, and capital expenditure therefore tends to be counter-productive, since every new transmitter or studio has to be staffed and maintained while bringing in no additional revenue. This is also the objection to large-scale borrowing to solve our financial problems. But of course, whether it brings in more revenue or not, capital has to be spent on extending and perfecting our services. In fact BBC 1 is now available to 99.5 per cent of the population of the United Kingdom, and BBC 2 to almost 80 per cent, and interference-free radio reception on VHF to 99 per cent of the population. We plan to improve on all these figures.

Of course the size of our income is governed not only by the amount of the licence fee, but by the number of licence holders. We have been fortunate in that, first with radio and later with television, we have had an automatic increase in income as more people bought radio or television sets. In the last ten years also the number of television licences has almost doubled, but saturation point is now approaching and the curve is flattening out. We hope that the supplementary £5 colour licence will to some extent compensate for this. In 1948 there were 46,000 black and white licences in force; four years later there were 1,400,000, a thirty-fold increase. Colour is now in roughly the same situation as black and white television in 1948, with 50,725 licences issued. With colour available on all three channels, the supplementary licence will bring in an increasing sum in the next few years.

The development of colour, an expensive process, has necessarily had to be paid for out of ordinary licence income. Now we are beginning to recoup on our investment. One day the BBC will no doubt be faced with the problem that the market for colour licences is saturated, and we shall have reached the kind of financial plateau we are now approaching in the case of black and white licences. But, given the present broadcasting framework I see nothing in the future, however far ahead I look, which would really financially threaten the licence fee system, for three reasons. First, there is the principle of the increasing number of licences, which is a hedge against rising costs. Second, where saturation is reached any licence fee increase would be closely related to general rises in wages or in the cost of living, and very possibly less than either. Third, in any large organisation, especially one which is largely technical, there are always continuing opportunities of reducing costs (e.g. automation, computerisation) which can make (and in the BBC definitely have made) a substantial contribution towards keeping the licence fee down. But that will be a problem for a future Director of Administration.

Conclusion

I have in this chapter briefly surveyed the various alternative methods of broadcasting finance as the BBC sees them. I have explained that we regard Pay-TV as unacceptable on the grounds that it is by its nature exclusive and socially unjust, that we consider that advertising on any of the BBC's services is incompatible with the preservation of the BBC's independence, and that the same objection applies to finance by Government subvention. I have also indicated that no other adequate sources of finance seem to exist, other than the licence fee. Finally, I have explained why we feel that this is in every way the best method of financing public service broadcasting, and that it is not, and not likely to become, a heavy burden on the ordinary viewer. Our policy may in fact be summarised as a belief in the licence fee, the whole licence fee, and nothing but the licence fee.

3

Broadcasting finance: some possibilities

Sir Sydney Caine

I wish to draw attention to the various ways in which broadcasting may be financed, and I do not want so much to defend any particular alternatives, but rather to speculate about possibilities. I will not, therefore, cover the same ground as Mr Arkell in the previous chapter, but I will attempt to indicate what seem to me to be the issues of principle involved, without necessarily saying how I think those issues should be resolved.

I want to look back further than Mr Arkell has done, and think what the possibilities would be if we were starting from scratch: that is, thinking about the replanning of the structure of broadcasting which might take place in the next seven years, on the assumption that it may be worthwhile thinking about comprehensive changes, and not merely planning the most convenient way forwards from where we are now. In this we want to make allowance for the possibilities, the technical changes, and other exciting advances referred to in Mr Stonehouse's chapter. The problem of financing broadcasting is to determine the method by which we may most conveniently raise the sum of money thought to be necessary.

I would like to make three points, all of which are, or may be, involved in methods of financing. First, the manifold purposes of which broadcasting has, and the need to allocate resources if all those purposes are to be achieved. Secondly, the need of the producer for all reasonable freedom of operation. And thirdly, participation, that blessed, if somewhat ambiguous word. By participation, I understand some means by which the ordinary member of the public, the ordinary viewer or listener, may play a larger part in the decisions about what we broadcast.

Any system of financing needs to be judged by whether it is likely to raise the funds needed for the full development of broadcasting; what are likely to be the external controls involved in it, and what is its influence on participation, in that sense. And I would like to go a little further on that third point, because I have considerable sympathy with the viewpoint that broadcasting is too important to be left to the broadcasters, and I would be glad to see the viewing and listening public able to have a larger say. It has been suggested that the concept of more democratic control might be realised through some form of elective approach. To my mind there is much to be said for a more direct form of democratic control, and for finding some device by which viewers and listeners can directly influence the types of programmes that we broadcast. A criterion in deciding between different kinds of broadcasting system must therefore be, how far each is capable of developing a mechanism for viewer participation of an effective kind.

I think too, that the interests of the producers, and those of the viewers are often opposed. A producer needs the freedom to devise his programmes as he sees fit, but it has always to be remembered that whether we are providing entertainment or education, the real test is whether the programme gets an audience. It may be satisfying to produce the right kind of programme, but this is immaterial if nobody looks at it. That the producer may be free, but know whether he is satisfying the public, we may need some devices other than those we have at present for connecting him with the public.

From the point of view of the economist, a system of charging, or raising revenue, or pricing, normally plays a part also in the process of deciding what is the appropriate sum to be paid for that service, who is to decide exactly what it consists of, who is to provide that service, and how much is to be enjoyed of that service by the individuals who pay that price or charge.

More simply, when you buy goods in a shop you are, by your choice of what goods you buy, helping to settle how much money in the aggregate will be spent on a particular kind of goods, how much of those goods you will consume, and which of various alternative producers will produce them. There are obviously great technical difficulties in applying an ordinary pricing system to broadcasting, as it is applied, say, to the comparable service provided by the Press. If it were possible, I feel that it would be a good thing if the consumer of broadcasting had more direct ways of making his preferences effective, by some comment—something analogous to *Which*, perhaps—and it is worth thinking seriously whether the system of financing should partake at all of the character of a normal pricing system. I gather that it would not shock the Postmaster General too much if in the future the amount of television, particularly hours of broadcasting, were settled by what the viewers and listeners were prepared to pay for, rather than by order of the Government. Perhaps this is one of the crucial issues, how far we want to see the ordinary individual playing a part in decisions, and how far we want them to be made by the Government or by some quite small group of people, the BBC Governors, the ITA Board or some other authority, settling things in what they believe to be the public interest, but not necessarily in the way in which the majority of people would want.

The choice of method of financing is therefore intimately bound up with the whole issue of the structure of the broadcasting system. Should it be in the hands of a single organisation, a Government agency, as in Communist countries; or in those of an independent, though Government-supported agency, as in this country before 1955? Or should the agencies responsible be multiple, but limited in number, as they have been in this country since 1955? Or should they be multiple, but unlimited in number, save by technical factors, as they are, in substance, in the USA? In practice, the role of the financial arrangements in the control of a broadcasting system is likely to be smallest where the basic system is monopolistic, and greatest, where there are the largest number of actual or potential broadcasting operators.

Let me then turn to the possible means of finance, here perhaps covering some of the same ground as Mr Arkell, but I hope usefully and differently. Broadcasting can be financed by the allocation of funds in the national budget, raised by general taxation; or by local authorities out of local taxes. Or it may be financed by the charging of a licence fee. Or it may be financed by the sale of advertising time.

And, leaving aside the minor sources of finance, there is the possibility of what is known as Pay-TV.

Any of these can be used in combination: licence fees can be supplemented by Government grants or, as sometimes happens, part of the licence income can be annexed by the Treasury. Advertising income can be used to supplement either licence fees or budgetary provision or, as in this country—and certainly it works quite well at present—one part of the system may be supported wholly by licence income, and another part by advertising income. Again, advertising income can be in part annexed by the Treasury, as is done on a large scale in Britain.

Examining these several methods individually, I am inclined to agree with Mr Arkell that direct support from the national budget need not detain us. It has not commanded much support in this country because it necessarily brings the broadcasting operation more under the direct or indirect control of Government, and exposes the broadcasting agencies much more to the danger of having the amount of their income varied according to the exigencies of the current economic situation. It therefore inhibits long-term planning. And it provides no semblance of a mechanism by which listener or viewer participation can be expressed through the financial structure.

Licence fees offer in practice a good deal more assurance of independence. They are normally fixed for substantial periods, and not subject to frequent debate and discussion like an item in the budget. But the fact that the licence fee *is* fixed by the Government ensures that the Government is free to make its conditions in fixing that fee, and in deciding how much of the proceeds would be handed on to the broadcasting agency. And Parliament could, if it wished, debate the licence fee. What has happened has been that the convention has grown up that these things do not happen. But the conventions could change. I think the editor and I, as being concerned, or I indirectly concerned, in university matters, have been aware that there is a convention which was formerly very strong, that Parliament did not debate university finance. This was dealt with through the independent agency of the University Grants Committee. That convention has considerably weakened; Parliament does now have a good deal of influence on university finance. The Controller and Auditor General is in future to have access to the books of the universities. I mention that only to emphasise that the conventions which have given independence to the system of financing university accounts are not quite sacrosanct; they are not written into the constitution of the country, and can be changed. Similarly, the amount of the licence fee, the proportion of it made available to the broadcasters, although at present governed by one set of conventions, can be influenced by financial considerations not directly appropriate to the broadcasting operation. As I think the BBC would readily agree, increased income is not always forthcoming as promptly as broadcasting considerations might seem to dictate. As for participation, the licence provides a somewhat rudimentary instrument for the expression of consumer choice, since people will presumably not take out a licence unless they think what they are going to see or hear is worth the price. But it is a very blunt instrument, since the viewer or listener pays for a whole year's viewing or listening as a single package, and has no means of expressing through his payment his choice of programmes, or of varying his payment according to the amount of his viewing or listening.

In the main, in this country, the licence system has clearly shown itself capable of raising the sums necessary to operate the BBC services: that is, the greater part of the total broadcasting activity of the country. And one could reasonably guess that it would be capable, if the Government were willing to face the political odium of increasing the licence fee, of raising substantially larger sums, perhaps enough to pay for the whole ITV operation as well. But it must be more problematic whether it could go further and raise all the money to finance further extensions of services, for example a fourth general channel.

There must always be some maximum level beyond which people will not think it worthwhile paying for a licence. So far we have been fortunate here, in that that level appears to have been well above that needed on a financial calculation; but this has not always been the experience of other countries, and may not continue to be so here.

Reliance on licence income obviously works best where all broadcasting is controlled by a single authority. If there is more than one agency, awkward problems of allocation must arise. It is, frankly, difficult to see the BBC and the ITA agreeing between themselves on a division of the spoils, if both had to be financed by licence revenue. It would be left to the Government to settle this, so increasing its power to interfere. It is possible for the fee income to be divided between a number of competing agencies according to the expressed preference of viewers, as I understand is done in the Netherlands; and that is a further step towards using the financial mechanism as a means of expressing consumer choice. But I do not know how satisfied the various agencies in Holland are with the resultant division of the income available. I suspect in any case that such a system could be expected to produce reasonably fair results only when a good deal more than just two competitors for public favour were involved.

Next I turn to advertising income. Many broadcasting agencies all over the world use this method. The ITV system here and the American broadcasting companies subsist on advertising. General experience is, that in the richer countries a fully adequate sum can be raised on the basis of a volume of advertising which is tolerable to the mass of the public, however much some people dislike the sight of advertising on the screen. Indeed where, as in the UK, the channels open to advertising are limited, the system tends to raise more than is needed on any reasonable basis to maintain the system. As is well known, because of this the commercial operation in this country has proved far more profitable than anybody had dreamt.

In the years 1956 to 1962, out of gross advertising income of £300 million, the companies were left with profits of £60 million after tax, the Exchequer taking in various forms of tax, over £70. Since 1962 the Exchequer has taken increasing toll of the advertising income, and in the last two years for which I have figures, the companies were left with profits, after tax, of £25 million out of gross advertising income of £170 million, the Exchequer taking, in various forms of tax and levy, a total of £77 million, or 45 per cent of the gross income.

In spite of these changes the profits remaining to the commercial TV companies are still quite generous, and generous enough to make the task of awarding programme contracts at periodic intervals when new contracts have to be awarded, an embarassing one for the ITA. I suspect that much more thought will need to be given to that process of awarding contracts before the next round takes place.

In other words, the problem of paying for television from advertising income, as currently operated in the UK, is that it produces too much income, that is, too much for the needs of a single television channel. If there were a second commercial channel the total advertising income would probably be not much higher, because the audience open to the advertisers would not be greatly increased; but it would still probably be enough to cover the total cost of two channels. Should the BBC ever accept advertising there would be a more significant addition to advertising income, because a larger audience would be available. But it would still be anybody's guess whether the increase in income would cover the BBC's costs as well. Basically, one of the troubles with advertising income is that it is determined by the competition of television with other media of advertising, not by anything inherent in the costs of television itself. It is a matter of hit-or-miss whether the resultant income corresponds closely with the costs of the broadcasting operation.

The advertising basis is obviously more independent of Government decisions than either licence fees or budgetary allocation.

Fears that it might be said to lead to control by advertising interests have been a good deal less justified in our experience, as the system has operated here. But there may still be legitimate fears of production budgets being kept too low by profit-seeking by commercial television companies. So far as consumer preference or public participation is concerned it does give a strong indirect influence to the views of the public, because advertising rates are sensitive to changes in the estimated viewing audience, and operators have therefore a strong incentive, some may think too strong an incentive, to put on the programmes which will attract the largest audience.

Lastly there is Pay-TV, the system under which, either by a coin-in-the-slot device or by some recording apparatus, the viewer pays for what he actually views. I myself attach no importance whatsoever to the experiments which have been carried out so far. That is, the firms concerned have not been operating in conditions which postulate the use of Pay-TV as the basic system of financing. The experiments have been small scale operations acting in competition with systems of financing by licence and by advertising revenue. I do not think therefore that we can compare the results of these experiments with what would happen if all broadcasting were financed on a Pay-TV basis. In particular, I have never been able myself to follow the argument that Pay-TV has to be exclusive. If a small scale experiment is trying to find its way into a system already dominated by great established interests, it has to try to offer something which those interests are not offering; that is, it has to that extent to be exclusive. But I cannot see why a general system of Pay-TV should be any more exclusive than the newspapers are exclusive.

If anybody can choose which programme he wants to look at, I do not see how that is any more exclusive than the present situation in which people choose between the three channels which we have available. So to me the argument of exclusivity is purely one relating to a particular situation in which Pay-TV is trying to establish itself against established systems. Therefore, if we are thinking in much more speculative terms of a combination of systems in which all broadcasting would take a part, there is no intrinsic reason why Pay-TV should not be combined with advertising, so enabling the charge to the viewer to be reduced, just as the price of newspapers is reduced by their large advertising income. Equally there is no reason why some programmes should not be provided on a Pay-TV basis free of advertising. People could pay more for these if they thought it worthwhile paying more to exclude the advertisements.

Pay-TV would certainly give the consumer the fullest opportunities of participating and of influencing the programme content. There are, I am told, very considerable unsolved technical problems in the way of establishing any universal system which would substitute Pay-TV for other systems of financing. But bearing in mind that it may in future be possible to introduce a general system of wired broadcasting, I think that the technical problems are not insuperable. It seems to me that if we are speculating about what we can do in the future and trying to review all the possibilities, Pay-TV deserves a lot more investigation—a lot more impartial investigation—than it has had hitherto.

To summarise, we may dismiss the alternative of direct budgetary financing for the reasons already given. Equally, there is not much reason to expect that local authority financing will play any significant part in broadcasting finance in this country. Financing by licence income is clearly capable of raising enough income for the operation of services on the scale of the present BBC operations, and might be capable of doing somewhat more: but it must be doubtful whether it can easily raise all the income that may be desirable for all the purposes of broadcasting in the future. It raises constant dangers of inadequacy of income because of Government reluctance, for political reasons, to raise the licence fee. There would, moreover, be great allocation problems if any attempt were made to finance two or more separate agencies from licence income. Finally, it does little to serve as a means of expression of consumer choice.

Advertising income, on the other hand, appears *prima facie* much more likely to yield an adequate income, but it is by no means certain that income from it would expand enough to cover the whole cost of all operations. In the present circumstances, when the income available greatly exceeds the cost of operation of the Independent Television service, the commercial attractiveness of contracts creates very considerable problems in the selection of contractors. This system does, however, serve as a powerful if indirect mechanism by which viewer participation can be brought to bear on programme decisions.

Pay-TV, finally, has many theoretical attractions, and if we are thinking of a comprehensive review, may be the best system to aim at as an ultimate objective, but probably more technical thinking about it is needed. If wired systems become available, Pay-TV may offer something much more flexible than the present system, inevitably limited by the small number of channels available. It could give scope for all kinds of experiments, including experiments with regional programmes, without the permanent commitment of resources involved in adaptation of the present structure. That indeed seems to be the most important criterion in the choice of allocating finance to the planning of systems—whether it opens the door as widely as possible to experimentation and new developments, and to participation by the public in what is still a rapidly developing technological field, or whether it acts as a drag on the growth of broadcasting.

4

Radio for local communities: a chairman's view

Sidney Holgate

I find it a little difficult to believe that there is any real justification for my contribution because it seems to me that in order to talk about localities, communities, and service, the real need is for a geographer, or at least a sociologist, and not for a mere mathematician to whom groups and sets have very different meaning. I have no professional broadcasting or journalistic expertise (I understand how 'mush' is produced only in the electromagnetic sense) and I do not think I am a very typical consumer, since I regard my television set primarily as a means of enabling me to watch test matches and international soccer matches that I am unable to visit, and I use my radio set for the incidental assimilation of information at times when most of my attention is concentrated on something more important like eating breakfast or writing a lecture. Perhaps, indeed, the only explanation for my contribution is the possession of a fairly thick skin of the kind acquired involuntarily over the last few months, while acting as an inadvertent buffer state between, on the one hand, the verve and enthusiasm of an energetic station staff, and on the other, an indeterminable audience, which is only occasionally irate, and whose needs, whatever that word means, are not very easy to judge.

However, I will try to consider the problem as a layman, and I think I ought to begin in the traditional scientific sense by trying to define the terms. The word 'community' is so loosely used in Britain, meaning anything from the groups which originally provoked the now almost non-existent tribal warfare between the Scots and the English, to the smallest of independent factions in the tiniest of societies, like the inveterate attenders at county cricket matches and symposia. It is all too easy to assume that a community consists of the population of a particular area, or possibly of the people who happen to work in a given industry, or perhaps of the people who happen to belong to a given class (if you know what that means) or to a group of people who have had a given type of education; but all these assumptions it seems to me are false. In this technological age, and in this day of increasing communication, to assume that there is any kind of division of the people into communities by area of residence or by nature of occupation, I regard almost as a contradiction in terms. No longer do people in any particular section of the country necessarily regard themselves as intrinsic to that area, and few of them indeed are really indigenous; and the facile assumption that, for example, North-East regional radio should concentrate on Blaydon Races or that Border television should devote a substantial part of its programme to Hadrian's Wall, or that Radio Stoke should talk about nothing but pottery seems to me to be much too easy a one to make and indeed to be almost self-evidently false.

Every area of this country now consists of an agglomeration of independent groups, an accumulation of minority sections and interests, dovetailed a little, overlapping often, in some cases integrated, in some cases adding up to a majority of the whole, but nevertheless fundamentally minorities, whose behaviour and interests are conditioned just as much by their particular work and leisure as by the area in which they happen to live.

If one looks at the question a little more closely, and tries to decide also what the word 'local' means, particularly in the sense of local radio and local television, the problem becomes even more complex. It is easy, as has been done in some cases, to draw on a map a particular area and say 'that is the area covered by your station', leading you to the rather ridiculous situation that we have in parts of County Durham (and we are supposed to be a County station and not a local station in the narrow sense of the word) in which, for instance, Hartlepool is technically not part of our service area, but is in fact embedded in the reception area and is surrounded by small villages and towns, all of whose population look to the Borough of Hartlepool for their leisure, their entertainment and very often for their employment. I do not believe one can define 'local' in this narrow sense, in the sense in which you would say that West Bridgford is not part of Nottingham, or that Salford is not part of Manchester, and one can think of a host of others. I wouldn't, however, like it to be thought that I am necessarily advocating regionalism rather than localism, because the two things are rather different. What I am suggesting is that within a given community there will be people who feel an affinity to one section of the country together with people who feel an affinity to another, and that geographical schism is wrong for broadcasting. There is, I believe, some doubt as to whether Stoke is part of the West Midlands or the southern part of the North West (the answer is of course both) and to attempt to draw rigid boundaries with mass media like radio and television does seem to me to be a little ridiculous. And I don't think the Maud Commission is going to solve this problem in any way. If the Commission in fact recommends that a particular area shall be an administrative area, I cannot see why this should necessarily coincide with the service area of any particular radio or television station just because you happen to have labelled it 'local' or 'regional' or whatever word you choose, and it won't coincide unless you can have a written guarantee that the boundaries will be impervious to transverse waves, so that only that area receives your transmissions. The day of the rigid boundary is gone, it remains little more than a postal or administrative convenience. It is absurd, uneconomic and unrealistic to attempt to define areas as tightly as county and borough boundaries would determine, at least for the purposes of broadcasting. I think the logical inference from this is that one has to think in terms of larger localities than have sometimes been thought of. It would seem to me to be absurd to try and have a local radio station in the Manchester area which, for example, excluded Oldham, or which pretended that some of the dormitory towns of northern Cheshire necessarily had different interests because they had a different postal address. At the same time there must be a limit to the size of a locality which is to be served by a radio station, and perhaps even by a television station, and it might be that the answer is a series of county radio stations comparable with the one we have in Durham, together with conurbation stations like Radio Merseyside and with area television stations like Tyne Tees or Border. But if this is the answer it would seem to me self evident that co-operation and not competition should be the order of the day in serving this kind of local community, unless the whole operation is to be economically unviable.

But it is easy to say 'serve' the local community; the real crux of the problem lies in deciding what you mean by 'serve'. The straightforward scientific meaning I suppose is 'meet the needs of the community', but it is not very easy to decide what the needs are, especially if you believe as I do, that the community in any case consists of an agglomeration of minority interests. Do you regard their need as the kind of things that they say they want, or would like to hear, or would like to see, and if so how do you decide what it is? You face the problem that one faces in almost any democratic community: that you cannot have a public referendum every time that there is a minor issue to be decided. And so if you attempt to assess the need, you tend only to hear the voice of a vociferous but tiny minority, just as is so apparent in the student world today. You will never get to know the opinion of the man who does not really understand what he needs, and certainly not of the people who do not care what they need, again whatever 'need' means. If you interpret 'need' as meaning what people say they would like, then your way of serving a particular community, at least as far as local radio is concerned, would be fairly simple. You spend 85 per cent of your time playing 'pop' music. On the other hand, one might take the view that it is the job of a sophisticated and enlightened group like us to make sure that the community gets what we think it ought to need—a sort of compulsory culture at the flick of a switch. I need not say how ridiculous it is for us to think that we have any right to take this kind of decison, or indeed use the mass media at our disposal in order to propagate this kind of idea. It is the job of broadcasting in my view not to be evangelically educative, but to stimulate the cultural requirements of its audience incidentally, by arousing their curiosity. And one might perhaps do this if one always maintains in view that the community is a collection of minorities. One may perhaps, say What is there particularly local about this? Why for example can you not stimulate interest in good music more effectively by a television broadcast of the last night of the Proms than by a local broadcast of a similar type of music played by a smaller and probably less efficient orchestra? But I think there is a difference. The man who has no interest in this kind of music is not likely to watch or listen to a national broadcast of it (unless it is the result of inheritance factor) but he might well be attracted to listening to a similar programme on the local station because he knows someone who is there; and in the process of admiring or denigrating the activities of the boy in the next street, he might well acquire some kind of interest in the subject matter itself. And this of course is by no means confined to music. This personal interest, this personal attraction of particular, known individuals is perhaps the greatest weapon in the hands of local radio and television in stimulating interest in various aspects of culture, sport and almost every other activity. Time after time one discovers, certainly in the student population, that a man develops a considerable interest in someone else's subject once he has tasted it, and he is only tempted to taste it because someone known to him personally is interested in it, because he goes to see his workmate's son perform in the local operatic group, and discovers that he likes Gilbert and Sullivan. He rubs his ear against the radio set out of vicarious pride and a bit of the broadcaster's mind rubs off on him.

I would not like it to be thought that I regard this as the only, or indeed as the prime, object of local broadcasting. I think it is our responsibility to some extent to give the public what we think they need, what we think will be useful to them in the widest sense of that word, what might make life a little more full for them, but we also have a serious responsibility to provide them with what they actually want even if privately we pity them for wanting it.

I believe that a very large number of the population, indeed the large majority of the population, do not want just local information and news, etc. I believe that they want a significant amount of news, information, sport, culture, etc. at a national level; and I believe at the same time that they need a certain amount of very local provision where their interest is much more personal. I do not believe for one moment that because a man lives in north-eastern England he is not interested in some of the things that happen in Devon and Cornwall. I am deliberately not going into details because that is the job of the professionals who make their contributions in other chapters, and I would not for one moment venture to trespass on territory which they know so very much better than I do, but I am sure that if any local or regional station in whatever medium is to be successful, it must carry a significant part of national news coverage, and of items of national interest, and of national programmes in every sphere that there is. And at the same time I think the area stations ought to be feeding things into the national network as some of the BBC regions do at the present time.

But the personal element must not be lost—I said earlier that I thought this was a question for co-operation and not competition, and I mean co-operation amongst all the methods of transmitting information, radio, television and newspapers. We really are not in competition with each other. The kind of pattern that I see for an event which obviously has a considerable local interest and a considerable personal interest for many people in the particular area would be pictorial coverage by a television report, necessarily restricted because the television station covers a pretty wide area; wider and yet more intimate coverage from an area radio station with personal elements introduced more frequently, normally longer than the television report, but still restricted in length and depth; and finally, although it would probably be simultaneously, a longer, perhaps less 'hot' but perhaps also more satisfying, critical appraisal in the local newspaper. That is the way it ought to work in practice if the local communities are to receive the best kind of service, but it will not work like this in practice unless there is co-ordination, unless the various services are integrated, and I do not put it completely beyond the bounds of possibility that there should be some co-operation between the BBC and Independent Companies in this area of influence.

And finally, if local broadcasting is to succeed, if it really is to serve the local community in the way that it should, then, although I hesitate to use what has almost become a dirty word in the university world, there must be participation. It is vital that the people who live and work in the area we are serving should take an active part in the day-to-day broadcasts from the station, should feel 'This is my station', not 'This is a station provided and run by someone else for the broadcasting of information and entertainment of their choice'; should feel that they are just as much a part of this organisation as they are a part of their own firm or tennis club. This of course raises by implication the ugly question of finance—if it is their station, do they pay for it—should local broadcasting be financed entirely from local sources, and if in fact it is so financed, perhaps from the rates, should the control of the station be in the hands of the people who are paying the piper? I have no doubt that some local financial support for a station is vital—I have equally no doubt that the station should not be entirely supported from local sources, and that it should certainly not be under the control of local authorities. I believe it is vital that such stations should retain a substantial measure of independence in the local sense, which means that they must not be run by local government. But I believe also that it is vital that they must be independent in the national sense, and not the mouthpieces of the metropolis, wherever the money comes from. The area stations must be locally controlled, as the present experimental local stations are. My council has received a great deal of help from London, but no instructions—we really are independent.

Whether we like it or not, the problem of interference (and I use the word strictly in its electrical connotation) will force us in the near future to a pattern of local or area broadcasting on VHF, both supplying local information and entertainment and relaying national programmes. These stations in this pattern must be intimate and personal rather than monolithic. We are moving into a new era in radio broadcasting; we are on the threshold of reformation, and we must not stumble over the step. All we have to ensure (but perhaps it is enough for the time being) is that we produce a progressive, professional, perhaps precocious, preferably prosperous infant, and not either a Victorian aunt or a Big Brother.

5

Radio for local communities: a manager's view

Kenneth Brown

Until recently an aspect of the history of broadcasting has been constant growth from smaller to larger units. Another has been a progression from the simple to the complex in techniques and methods of working. While this latter aspect has been particularly noticeable in television, radio techniques have become more complex too. The amount of time spent in preparing an hour of broadcasting has been increasing and demarcations have tended to be very rigid. At the same time, as equipment has become increasingly sophisticated, it has become more dependable and simpler to operate. Taking advantage of this and impelled by a categorical imperative that radio must become cheaper to produce, local radio has deliberately faced and solved the urgent problem of simplifying radio techniques.

I am reminded of the social anthropologist who was studying the effect of civilisation on some South American Indians. He noticed that, from time to time, the women of the tribe addressed a tree in the middle of the village and thereafter appeared to be listening to it. He asked what they were doing and they explained that it was the instrument with which they communicated with their menfolk who were working in the nearby town. They had chosen a tree, they said, as their method of communication. 'Why did you choose a tree?' asked the anthropologist. 'Because we are poor' they replied. 'Of course, if we were rich, we would use the telephone.' 'But why use a complicated telephone when a simple tree will do?' the anthropologist might have asked. I feel that the people who designed the working structures of the BBC's new local radio stations must have asked themselves the same sort of question. When I look back during my years at the BBC on the complex production methods which have grown up during the years and the comparatively lavish staffs which have operated them, I now feel that we should have been looking for simpler and cheaper methods many years ago.

Of course, with the greater complication of network and regional radio, extreme professionalism is devoted to each minute aspect of production and in comparison I often find local radio too hurried, too little prepared and rather untidy at the edges. But like the Indians' tree, it is designed for simpler more direct kinds of communication, and covers smaller areas. And unlike the telephone, it is, or should be, organic and rooted in the community in which it grows.

The simplified structure of local radio is a consequence of the basic fact that if we are to have local radio throughout the country, we shall have to have many stations. However you define the model community which such a station would serve, this means a great many more radio stations than we have now. If the nation is to be able to afford this multiplication of stations—regardless of how we finance them—we must be able to run them in a much cheaper manner than we run conventional radio. Simplicity is the only way.

The principle behind professional radio so far has been expertise within a relatively narrow range. Even the simple radio production, say a daily current affairs news magazine, demands a producer or producers, a secretary, one operator who controls the various sources of sound from the control panel, sometimes another one who operates the tape machine and edits the tapes before transmission, an announcer or 'continuity assistant' who announces the programme before it comes on the air, a 'narrator' or 'anchor man', and finally the freelance reporters who go out with their portable tape recorders or who interview people in the studio. Local radio realised from the start that all these tasks, while being responsible, were also simple. And that many could be combined. The producer, the control panel operator, the narrator and the interviewer, for example, could all be the same person. We realised that intelligent, well educated young people, equipped with reasonable manual dexterity and good physical and mental co-ordination were capable of undertaking a combination of the tasks which we had traditionally expected of radio producers, equipment operators, reporters and sub-editors.

In fact, apart from the specialised task of maintaining the technical equipment, the local radio man is capable of combining all the skills necessary for producing all the station's programmes.

In a sentence, then, the programme staff of a local radio station act as interviewers, news gatherers, sub-editors and producers. They edit their own tapes, record their own programmes and introduce and narrate their programmes while operating the control panel at the same time.

Productivity

It is unnecessary to point out that this interchangeability of functions leads to a marked increase in the productivity of radio. The main task before us is to minimise staff and maximise output. In order to maintain standards we must be careful not to accept this aim too uncritically. As soon as one becomes cost conscious in broadcasting there is a danger of worshipping cost at the expense of value. The daily output of the BBC's present eight experimental radio stations varies, according to the outlook of the station manager and the size of the area covered from as little as four to as much as eight hours a day. The average number of staff at the stations is twenty-one. It is difficult to make comparisons of productivity between these stations and regional ones. Comparisons between programme budgets are perhaps more easy to draw. For instance, the total programme budget of my station is about £300 a week. A typical simple news magazine produced by a Region, on the other hand, running for twenty minutes a day, five days a week, might cost, including payments for news and freelance reporters and line charges, as much as £250.

Comparisons are oversimplifications of course, but £250 for one hundred *minutes* as against £300 for thirty to thirty-five *hours* is startling in itself. A further comparison—which I would not push too far— but it is worth mentioning, is that the cost per hour of network radio in 1966 was over £500, and that our first rough costings of the cost of local radio is between £30 and £40 per hour.

The very conditions laid down, quite deliberately, by the BBC for the local radio experiment—a small staff and little money—is a stimulating and I may say a most unusual exercise in management for one who, like me, has come from regional radio and will shortly be returning to it. The local station manager is beset with the problem of productivity, largely in terms of use and deployment of staff to an extent which I have never experienced before. To set against this, working within a centrally agreed framework of staff, equipment and programme money, I am freer in my programme decisions than I have ever been before in the BBC. This brings me to the next fundamental aspect of the structure of local radio—its flexibility and diversity.

Flexibility

Regional radio in the BBC has always had a degree of independence. But it is, after all, part of a large integrated network and not only does it produce its own programmes, it also devotes large resources to producing programmes for the network itself. Thus decisions in broad terms on the amount of broadcasting it produces and the times at which it broadcasts are heavily influenced by the needs of the London network of which it is a part.

The local radio station is part of no network and it is no part of its task to serve a network. This fact alone gives the station manager extreme freedom and flexibility in his programming Sadly, however, it is not the whole story. If the local stations could produce their own programmes all day long, if they could give their listeners a genuine choice of programmes all day long, even if for several periods of the day such a choice were only programmes of gramophone records, the local radio challenge would be much sharper. This is not the case in this country because of the copyright laws. Because of these laws we cannot fill in the intervals between our own programmes with records, as is done in the United States. Like our regional colleagues, therefore, we have to fall back on the London networks for what we call 'sustaining material'. With this difference. We broadcast our own programmes exactly when we like and we use as sustaining material whichever network we like, choosing whichever network happens to suit us at any particular time of the day. I ought to add also, that we have, naturally, the ability to interrupt either our own programmes or those of the sustaining network of the moment at any time according to circumstances, as we, and we alone, decide. Compare, for a moment, this flexibility with the inflexibility of the networks and the regions. Because of the large areas they serve and the rigidity of the network pattern information must be generalised in its nature and can only come at set times. If Kruschev is dismissed, you may well get a news flash during a 'natural break' but if the main road is flooded on your route to work in the morning the only station that can tell you—and it can tell you straight away—is your local radio station. And at that particular moment there seems to me no doubt as to which message is the most urgent to you.

It seems to me that today the country is not only in crisis nationally but also, although less dramatically, locally. The more complex society becomes, the more difficult life becomes at the local level. Local radio, given the flexible pattern which it has, can exploit this situation. I do not think we have yet done so fully, for reasons which I shall later explain.

But first let me tell you an illustrative story of local radio where it is already well established and accepted. It happened to a Durham man of my acquaintance who was in Australia on business and who arrived at his destination, a seaside town, the day before he was expected. No one in the town knew him by sight and he decided, after he had booked into his hotel, to spend the day on the beach. Lying in the sun, he was listening on his transistor to the local station when he was astonished to hear the disk jockey say' 'Will Mr Brett (which was his name) please make himself known to the beach attendant who has a message for him.' (He probably said it, by the way, in a much more friendly and less official way than that.) Intrigued (how on earth did anyone even know he was in the town, let alone on the beach?) Mr. Brett went to find the attendant, who handed him his wallet, which he had apparently dropped on the beach. It had been handed to the attendant, who had opened it, found out his name and 'phoned the radio station.

That story illustrates a flexibility which we share with local radio stations all over the world. But it raises wider issues. It raises the issue of service to the community. Radio has so far served the community in a general rather than a specific sense, because of the wide areas which it serves. I said a moment ago that society is in crisis at the local level. Perhaps I use too strong a term. Telephone lines break down, there are pile ups on the main road, huge traffic jams at holiday times, fog emergencies, power cuts. At another level there are small administrative injustices too localised and not sufficiently dramatic for the notice of the larger agencies of press or radio. In the local radio station there is now an agency available which can deal with these issues quickly and effectively. Take the unexpected blizzards in Brighton last year. They were no more than a news item for the national bulletins and no doubt they devoted two or three interviews to them on the South-West programmes. But on Radio Brighton there was a continuous information service in collaboration with the police until the emergency was over. During the floods at Leicester, the life of one old age pensioner may have been saved because she was listening to Radio Leicester's minute-by-minute reports of the flood's progress.

I don't think, as I said earlier, that we have exploited this aspect of our flexibility to the full. Of course, in real emergency, the local station comes into its own—to such an extent that Anne Duchene called us 'Radio Crisis'. This is because the local authorities at such times desperately need help and turn to every agency which can give it. But we can help the authorities and the community at a lower, less dramatic, but still important level. Here we depend on the awareness of the authorities that we can help.

At Radio Durham, we look for the tiny announcements tucked away among the classified advertisements of the local press which tell readers of forthcoming temporary interruptions of their electricity supplies. We then broadcast reminders to our listeners near to the time of the supply cuts. We have to take the information from the papers, in fact, because the Electricity Board refuses to tell us direct. We want in addition to be able to tell listeners about emergency cuts and indeed listeners have rang us up from time to time asking us if we can tell them when supplies will be restored. We are in negotiation with the Electricity Board about a service for listeners on those lines but the going is slow. In general, the attitude of the authorities—the police excepted—to our proferred help is one of reluctance to make use of it. Perhaps we are too new. But I am sure that when there are electricity cuts in the Australian town Mr. Brett visited, the transistors carry the news immediately on the local wavelength.

In other words, the uses of local radio as a community service do not depend solely on the willingness of the local station to help. They depend on the willingness of society to be helped. There is no doubt that a new community service having been created, the community will make full use of it sooner or later. My experience in the short five months Radio Durham has been on the air is that the public are more aware of our uses than the authorities. But different areas will have differing responses and other local managers may have different experiences.

Because the local station is physically nearer to its listeners there can be closer contact. If the station is in the middle of a large town, as at Leeds, the public can drop in to broadcast their complaints or queries. They can and do ring up—and in many amusing ways. Two, possibly trivial, examples from last week at Radio Durham. A mother rings us up at breakfast time, 'Will we please play 'Little Arrows' within five minutes'. Why so soon? She explains that she cannot get her son out of bed. 'Little Arrows' is his favourite record and if she tells him it is on the radio he will be sure to get up to listen to it. Also at breakfast time, a little girl rings up. She explains that she has forgotten to send her father a birthday card. Today is his birthday and she has just realised it. Will we broadcast her birthday greetings and perhaps he will forgive her. Or the people who rang *us* to ask for information about a recent electricity cut rather than ringing the Electricity Board. The examples may be trivial. But they indicate an attitude to a radio station which I think is new in this country. I remember giving a press conference about Radio Durham before we went on the air and after my definitions of our aims a journalist accused me of wanting to set up another kind of 'Citizen's Advice Bureau'!

On reflection I think he had a better point than I realised at the time.

Participation

But it is not sufficient if the radio station merely establishes this new close relation with its listeners which I have just described. It is a relationship which I imagine the best local newspapers have with their readers. Not only must the listener look to the local station as a steady source of reliable information which is specifically local in character, he must be encouraged to participate. By this I do not mean merely given the opportunity to broadcast. It follows in a local radio context that far more people in the area will broadcast than ever before. In a station like Leeds, for instance, situated in the centre of a densely populated city, it is claimed that at least one hundred people a day broadcast live from the studio. Locally, where there are five, six, seven hours of broadcasting a day and more, the act of broadcasting is going to lose its mystique and become something that most people will take for granted—something far easier for the majority of people than writing a letter to the local paper. By participation, however, I mean the more important function of local organisations in promoting, arranging and actually producing their own broadcasts. This is, of course, breaking down barriers in a way practically unknown in this country up to now.

I am in danger of repeating myself. But I must emphasise that local radio is simple radio and there is nothing intrinsically difficult about it. Our teams are predominantly young men and women who were not in the BBC before we started the new stations. My colleagues joined me in mid-January this year and had a mere five weeks training. One of our best assistants joined us as a newly graduated student in search of work during the vacation and has never had any formal training. He still has a lot to learn. But he knows more of tape machines and the operation of control panels than I do, and the number of microphone hours he has accumulated since June is more than I have registered in eighteen years at the BBC. Once a week, a youngster from a local gramophone shop in Durham comes to the studios and broadcasts his own forty-five minute record show. The University students produce their own weekly magazine during term time. Local teachers help to produce education programmes on behalf of the Education Authority. The Durham Consumer Group plans and produces six short programmes a week on local goods and services.

Professionalism

I can imagine the thoughts of some of my regional radio colleagues as they hear all this. They might well be asking what about standards ? what about professionalism ? And they would be right to ask such questions. I wonder if you have ever heard the remark which is said to have been made by Mr Sydney Bernstein of Granada TV ? 'Look at the BBC' he is supposed to have said, 'They do things properly, but their ratings are no better.' Now that remark may have been apocryphal, it was certainly cynical and I do not subscribe to it, in the sense in which it was meant. But it can be looked at in several ways. Professionalism, as we understand it in the BBC, tends to increase in direct ratio to the size of the area you cover, and the distance so to speak from your audience. The more 'distant' you are, the more slick, quick and efficient you must be. The listener is not on your side. But if you are a local station, if you rightly belong to the community, if you are 'their' station, the world does not collapse if you make a mistake. If—heaven forbid—you put on a gramophone record at the wrong speed, you might seem all the more human and approachable for doing it (as long as it does not happen too often). It's happened twice at Radio Durham since July and I hope it won't happen again. But we were more worried by our lapse than our listeners were. We need all the professionalism we can achieve and which we can learn from our regional colleagues. But not at the expense of the solemnity of Radio Four and the artificial highly 'professional' palliness of Radio One. I think, basically, relevance to the audience is what matters and in this respect, the smaller the area your station covers, the more relevant you can be. I think we can evolve a new professionalism of our own, but I do not regret that we shall always be somewhat more rough and ready than the networks. I think there are compensations—

in informality, friendliness and warmth that can make up for what we lose on the sheer professional level.

Finance

One of the objects of the present experiment was to find out to what extent local communities will contribute towards the running costs. At Radio Durham the wolf has been at the door so often that we have asked it in and domesticated it. I do not think that local contributions are a viable way of financing broadcasting. But they are an indication of grass roots support. It is heartening for instance that in County Durham, where no engagements were entered into before the station was announced, we have had contributions from the County Council, from Peterlee Development Corporation, from the Northern Arts Association and from the Diocese. It is not, I think, the actual amounts which are important. It is rather the token recognition that we belong to the community and are not thought of as a branch of a London organisation. But such friendly gestures will not, I am afraid, provide a permanent and satisfactory financial structure for local radio.

Personally, I stand by the licence fee as the only ultimately satisfactory way of financing public service broadcasting. Local radio is remarkably cheap by comparison with any other medium of mass communication. We ought to be able to afford it.

But we must be realistic. Suppose the politicians are not prepared to put, say, five shillings on the licence fee and suppose the BBC cannot find the money to finance the new stations from its existing resources? Perhaps I might quote the well-known financier who used to address his Board for two hours on his policies and who solemnly concluded 'Gentlemen. . . . These are my principles. But if you don't like them, I have others.' I think, and I speak for myself in this, that it may be necessary to consider the possibility of accepting advertising as one source, although possibly not the only source of revenue for running the stations. It will mean larger staffs in the shape of advertising departments, it will probably mean that people who now broadcast for nothing will, like the journalists who supply the news, demand a fee.

At the moment on local radio we paraphrase C. P. Scott and say, 'Comment is cheap, news is expensive.' When I address a non-commercial organisation, like a Rotary Club, I neither ask for nor am I offered a fee. That is at present the attitude at a non-profit making local station. But if we enter the business field, people's attitudes may alter. They may demand a fee. So advertising may make local radio a more expensive proposition. I do not think it would affect our independence any more than have contributions from local authorities. It might to some extent affect the amount of time we devote to minorities and it could affect the shape of our programming. And none of these things would I relish. But I hope that the BBC would not stick so rigidly to its principles that it would throw the baby out with the bathwater.

In any event, if a local station covers a small area, if it broadcasts on VHF only, if it continues to face savage restrictions on the use of gramophone records, is it a proposition which would attract a business man who wants more than a very modest return on his capital? These days, even the small television stations work on comparatively small margins. Finance apart, however, there is now no doubt that local radio works, fulfills a need and is appreciated. The case has been made. Three further questions remain. Will it remain under the control of the BBC, whether run commercially or by other means? What effect will local radio have on the BBC Regions? What is the ideal size of the area which the local station should cover?

Size

I am not equipped to answer the first two questions. In answer to the third, I think that the smaller and more densely populated the area the station covers, the greater will be its impact and its value. Radio Hull, Radio Brighouse, Radio Preston, would have been fascinating experiments. Radio Sheffield and Radio Leeds, nevertheless, cover recognisable communities and attract a large number of listeners within relatively small densely populated transmission areas. But to cover the country adequately with stations of this size, we could need as many as 150 stations. I doubt whether this is possible either technically, financially or in terms of adequate staffing. I believe that we shall have to think in terms of larger 'conurbation' stations, which will be more difficult to run and which will have less impact than the smaller stations. I also think that the number of stations and the size of the area they will cover will be related to the forthcoming recommendations of the Royal Commission on Local Government.

Many well informed people anticipate forty or fifty new areas of local government—city regions is one term for them, and it would seem logical that each such area should have its own radio station. Just how 'local' they will be remains to be seen. But certainly they could retain all the characteristics of simplicity, flexibility, diversity and participation which I have described.

If such a situation does arise, stations like Radio Durham and, even more so, Radio Merseyside, which cover much larger areas than were originally contemplated for so-called 'local' stations, but with roughly the same staff and resources, gain an importance beyond the mere proving or disproving of the original case for local radio.

6

Television for local communities

James Bredin

I propose to deal specifically with Border Television and its area, not out of immodesty, nor out of a desire to grasp a public relations opportunity for my own company, but in order to illustrate my argument by reference to a specific case of a television service in a local community.

I do not propose to discuss the problems of Border—I assume that half my audience would not be interested and the other half would be glad to hear that I was getting what was coming to me—but I propose to talk about what we do, how we do it and why.

Border has an area which is geographically large but thinly populated. It stretches from Stranraer on the west coast to Berwick-on-Tweed on the east coast, through Dumfries on the Scottish side, Carlisle on the English side and the Border towns, again on the Scottish side, of Kelso, Hawick, Galashiels and Jedburgh. The area takes in Cumberland and Westmorland and the Isle of Man. Its population is 592,000, of which the English make up 57 per cent, the Scots 35 per cent and Manxmen 8 per cent. When Border Television was set up seven years ago it was thought by most of the existing Companies not to be viable. The assumption was that one could not run an ITV station with a population of less than a million. My predecessor, in Border's first three years, proved them wrong. On issued capital of £125,000 he broke even in three years. Since then, Border has made a profit in each year, and in 1968 expects to make a profit before tax of about £100,000.

All that by way of background. What do we do by way of serving the local community?

First, we take seriously a part of the country which would be taken less seriously, to put it mildly, if Border Television did not exist. The area had not been taken seriously before we began to do so. The Scottish part of our area is not taken seriously by

Glasgow and Edinburgh because it is as far south as you can go in Scotland. The English part of our area is as far north as you can go in England. And the Isle of Man is—the Isle of Man. We take all three seriously. Border, like most regional television stations in ITV, does a daily half-hour news and current affairs programme called *news and lookaround.* We regard this as much the most important programme we do. It happens to be very successful, if I may be forgiven for being immodest about it. It is, in fact, the most successful of all regional programmes of its kind, regularly getting 70 per cent of the audience which can choose between BBC and ITV when it is transmitted. It could be an introverted local programme dealing with local trivialities. It is not. It covers events and opinion in the area without being condescending on the one hand or indecently parochial on the other.

Secondly, we work with other regional television stations in order to relate our interests to theirs. By this I mean that we regularly exchange material with our neighbours Tyne Tees Television and Scottish Television, with Granada and Yorkshire occasionally. Their equivalent of *News and Lookaround* may well be dealing with the same news story or feature in their terms. Our feature becomes less parochial and better understood when related to the same feature as experienced elsewhere.

Thirdly, we do the kind of programme that anyone can do; the after-dinner-conversation type programme. But we do it without, again, being parochial about it. *Border Forum* has local and national speakers in it, together with, as chairman, our own staff man who is already familiar to viewers and who can, as Dr Holgate has suggested, persuade people to become interested in subject matter which might well not interest them when it is offered from London. In the programme, earlier this year, we persuaded Cecil Day Lewis to talk about poetry and the Poet Laureateship. Thus it had its effect on more people than another kind of programme on poetry would have had.

Fourthly, we offer a platform from which our area can express itself. For example the Border Area has recently been engaged in a fight which was comparable to that about the proposed Stansted airport. The Ministry of Transport had decided to close down the railway line—known as the Waverley Line—which runs from Carlisle to Edinburgh through the Border towns. The case for and against closure was argued, and reported by us blow by blow, for twelve months. As reporters of this argument we, like everyone else, found it extremely easy to state the case for keeping the line open and next to impossible to get the facts which led to the decision to close the line. I therefore invited the Minister of Transport to give the facts in a half-hour programme in which he was confronted by the people who needed to have the line kept open. In the event the decision was not rescinded, but Border Television was the means whereby a local community was enabled to express itself effectively and to demand to be taken notice of.

The four conclusions I draw from all this are: (1) It is possible to serve a local community effectively, mixed though it may be even in terms of nationality; (2) It is possible to do so without going bankrupt in the process; and (3) It is possible to justify the existence of a station the size of Border, if it regards its primary aim as that of serving its area with programmes for that area. Its primary aim should not be to do programmes only about its own area. I agree with Dr Holgate that there is no reason why Border should devote its locally originated programmes mainly to material about Hadrian's Wall. Nor should Border fool itself that it is entitled to get its programmes networked simply by reason of its existence. If it gets a programme on the network, that is incidental. It is not the object of the exercise.

Our job, I believe, is to foster local pride and self-awareness without pandering to local triviality. That is done by dealing with local matters intelligently and by relating them to matters national and even international so that they cease to be trivial.

My final conclusion (4) is that staffing is bound to be a problem. A large turnover of staff must be expected and accepted because a small station is bound to be a training ground for larger companies and for the BBC. The answer, particularly on the creative side, is to go for young people with potential, give them a job beyond their present expectations and not to expect them to stay for more than a couple of years or so. When they are ready to go, the last thing you should do is to stop them.

Finally, Border does not delude itself that it knows what are the needs of the people who live in its area. As Dr Holgate says, any community consists of an agglomeration of minority interests. I put it another way—it is not quite an original thought—I wish it were. There is no such thing as the average viewer. There is only you and me and we are both peculiar. As long as we remember this, there is some chance that we can be of service to a local—or any other—community.

7
Regional devolution and standards of excellence

Alasdair Milne

I hope I may be forgiven for writing largely in terms of Scotland. I do know something of the operations of the three English BBC regions, and of what goes on in Wales and Northern Ireland and I will refer to them, but naturally I will make the greatest sense when speaking of Scotland, a nation of five million people with a long history, a long memory of that history and a deep-rooted culture of its own; in that sense, never a region in the normal sense of the word and certainly a people who have clamoured, increasingly over the years, for their broadcasting to be genuinely Scottish-based.

Because we have been at the business of broadcasting for many years, I think it would be helpful if I remind the reader very briefly of the developments of broadcasting in Scotland since its beginnings. For us, it began with radio, and, as it now seems to be developing—with local radio; before the Corporation was created, while we were still the British Broadcasting Company, radio sources were established in Glasgow, Edinburgh, Aberdeen and Dundee. They started with only limited output; but as the transmitter pattern advanced across the country, the output increased through the late 1920's and 1930's. As early as 1923 we had done our first programme in Gaelic, a year later our first experimental schools programme. By 1935 the BBC Scottish Orchestra had been formed and programmes of all kinds were being produced in Scotland. The Scottish Home Service—an amalgam of network and Scottish programmes—began after the war and the vigour of that output is still fresh in many minds. As with the North Region with its renowned school of radio feature producers, the BBC could feel that Scottish broadcasting was flourishing. If the picture changed, it changed with the arrival of television.

When television restarted after the war, the tendency undoubtedly was to concentrate resources in London. One must remember there were harsh restrictions on capital investment at the time and a limited licence fee; but it is fair to say also that few people at the top of the BBC then believed television had much future and, it should be said, there was then no hint of competition. It is hard not to feel that outside London we have been suffering ever since from that unwillingness to invest. In Scotland, we began in television in 1952 with an outside broadcast unit; there followed in 1955 a small presentation studio, in 1957 the renting of an outside studio capable of drama production on a drive-in basis and in 1964 the most modern drama-size BBC studio outside London. Throughout that period there was a growing development on the film and servicing side. At the end of 1968 we had eight film crews and associated cutting rooms; the drama-size studio I referred to; a small talks studio, a small studio in Aberdeen and a one-camera source in Edinburgh, an OB unit and full supporting services in terms of design, wardrobe, etc. We have substantial plans for full colourisation and further studio development. We are better off, unquestionably, than most BBC centres outside London.

I do not wish to bore the reader with further statistics or historical recapitulation; but something else I must add. During this period, the constitutional position of Scottish broadcasting underwent two significant changes: Beveridge in 1952 recommended that the National Broadcasting Council for Scotland should have authority over the content of the Scottish Home Service on radio; Pilkington in the early 1960's recommended that the Council should have the same powers over television programmes originated in Scotland. To give effect to those powers the BBC established a five-year plan for programme development of television in Scotland, to run from 1965–70. That development was frozen for two years as a contribution to the BBC's overall financial cuts—it starts again in the autumn of 1969. These then are the facts of the situation in Scotland: we are lucky—the English regions are not so well equipped in terms of facilities or finance; but then, as I said earlier, the BBC's thinking in the television age, always related to an uncertain financial position, has differed from that of ITV.

Before I approach the central questions that face any broadcaster working outside London, I wish in passing to make a purely personal point. I have spent most of my time working for the BBC but I have also worked for a number of ITV companies; and I do not go along with the belief that the federal structure of ITV, the existence of a number of small stations, necessarily contributes to the good of broadcasting by sheer force of variety. It might well do so—but we have not yet seen the evidence. Undoubtedly, as Professor Wedell says in his book *Broadcasting and Public Policy*, 'particularly in the smaller ITV regions, the television companies have added substantially to the interest of the local scene'. Of course a TV centre in Carlisle, Aberdeen or Plymouth must be a new focal point of interest, and most of the small ITV companies have worked hard and successfully to identify themselves with their own areas. But for the first ten years of ITV's existence, they seldom had much impact on the *general* broadcasting scene—there were only Anglia's plays and natural history series, Southern's children's programmes, the odd documentary from the late TWW or Tyne Tees, an occasional adult education series from wherever (if I offend by omission, I apologise). Mostly, it was the stranglehold on airtime of the network companies which told, and though the small companies were famed and successful in their own areas, I think they had far less chance to contribute to the national broadcasting picture than did the BBC regions—despite the BBC's centralised character. There are signs (or until very recently there were) that the smaller ITV companies would have greater chance of access to the network; but the assumption that the federal system, as we have seen it in action is *per se* a better force for the general improvement of broadcasting seems to me not proven.

Having got that off my chest, I have no hesitation in saying that there is not sufficient regional contribution to network on both sides of the fence—and I mean by that, creative television of a high professional standard; I take local news programmes and the like for granted. If I offend my friends at the Television Centre by saying so, too bad. I think I know the pressures under which they operate; the facts of a virtually static income in an inflationary world; the facts of greater efficiency in the throughput of programmes at the centre and, the reverse of that coin, high costs in many regional centres—though that of course is largely explained by lack of revenue preventing the use of capital equipment to its full advantage; the facts again of a competitive system and the gravitation to London of many of the artists and writers, with the result that it often seems (certainly from London) simple lunacy to risk a doubtful regional product against the proven efficiency and success of the London creation. And, incidentally, if any here scoff at our interest in competition, let me remind you that many of my colleagues, including both of us here, fought our way from a ratio of 70/30 against us to parity and, I am happy to say, the BBC's recent lead. Had that not happened, there are many voices in the land who might have demanded, and voted for, the dismantling of a public service system of broadcasting and its demotion to some kind of television Third Programme (for which I, for one, would not have worked). All these facts my London colleagues face.

And, of course, we live in times when the pull of London has long been a painful sociological fact; it is not so long since I heard a Minister of the Crown express the view that if people living in the Western Isles or the Northern Isles really wanted to stay there, then they had better stop complaining about higher prices and poorer living, or be sensible and pack it in and move to the mainland of Britain. London and the South-East pull people in—the provincial theatres struggle and then die and the actors move south; the agents move with them; the film studios are around London, London can provide more opportunities in journalism; salaries and money are higher in London. And yet for many there is an increasing distaste for London life and values—or what, to be fair, are *thought* to be London values; and with it a feeling that few in London understand or care about what happens beyond Barnet, Basingstoke, Brighton or Brentford. And backing such feeling powerfully, there is the Government's repeated pressure away from the south and towards development of the regions.

Enough of the background. What do I want to see? Sitting in Scotland, do I want more money? Yes, of course I do—though I hasten to reiterate we are not badly off. Have I got enough money to make real changes in television in Scotland?

Yes, I think I have but I would not wish to see our planned expansion stop in three years' time. The money we will have enables us to move forward from a solid base of (I think) good coverage in news and current affairs, good output in sport and religion, in schools and light entertainment, to extend far more widely than hitherto into the drama field, into increased experimentation in entertainment programmes, to make a beginning in children's programmes. Part of this development must be for ourselves in Scotland—we can and do for instance run an inter-burgh quiz programme which is fascinating to us in revealing our own ignorance or expertise in our history, geography, dialects and so forth—but is of no interest beyond the Border. There is a vein of Glasgow comedy which is accepted, albeit sometimes with reluctance, by the rest of Scotland and is highly entertaining to us; not to anybody else. The children's programmes from the network of both channels are extremely good but for us they leave great gaps which we must fill; and as in comedy, so with drama there are areas of Scottish life to which we ought to give dramatic expression, quite apart from our duty to encourage writers to live and eat in Scotland and actors to stay and even come home again. So more we must do, and in a wide range of programmes.

Then, standards of excellence—part of the title of this chapter and in many ways, the crux. Do you, as the Americans most certainly do in many of their local stations, accept that local originations are of a lower standard than the network product, and hope that the local community will take a charitable pleasure in the fact that it is done at all while winking at low standards? It happens here, too—'let's fill half-an-hour with a cheap programme of chat and we'll get away with it because it's our own.' I do not believe in this approach, though I admit to avoid it entirely is easier said than done. All of us in the BBC outside London and in non-network ITV companies operate on smaller budgets than the London producers do. One would wish it were not so—London-held theories about it being cheaper to produce in the regions are frequently bogus—but then there is not a bottomless purse to dip into. So we have smaller budgets on the whole; but one then has to ask the producer to work that much harder to ensure his work is up to the standards of production accepted throughout the nation. The audience is by now extremely sophisticated and will not, rightly, tolerate shoddy production, quite apart from its effects on our own professional self-respect.

No lowering of standards in local television, therefore, is our aim, even if the money is sometimes contemptibly small. We do not always achieve this but it is our continuing ideal. We also strive continually to achieve network showing. Here again our situation is different from, say, Yorkshire's—they are a network company, competing for certain areas of time; we have, traditionally, contributed somewhere around one and a half to two hours a week to network of all kinds of programmes and that is accepted as the norm. It must now be a norm from which we move forward; people in Scotland want more Scottish originated television but they rightly want it to be of a standard that can match up to anything in Britain. And that seems to me the ideal for a genuinely regional station—more of your own and if you make it well enough, a percentage is bound to be of national interest. If we cannot increase our contribution to both BBC channels—in drama occasionally, in light entertainment increasingly, in documentary programmes for the first time—then I will feel I have failed.

And perhaps at this point, I should touch on a highly contentious point. Voices are frequently heard in Scotland, upbraiding me for what they call 'my servile submission to my London masters', exhorting me to remember the ways of my fathers and stand up and fight for the independence of Scottish broadcasting. I have explained that through the constitutional position of the Broadcasting Council for Scotland, we have a very real measure of independence: the contentious question is—would we be better off if we were totally independent, living off our own Scottish licence fee?

Well, firstly, the sums are certainly hard to disentangle—no easier to sort out than are the economics and finances of England and Scotland in the present constitutional debate. There is no question that we could survive—though, for example, who would finance the extremely costly UHF transmitter development necessary for 625-line dissemination in our terrain? In some areas, I think we would be better off: in greater freedom, perhaps, to plan our own building and studio developments, in more flexibility in deploying staff without having to pay regard to current practice in London and other parts of the kingdom. In other ways, I think we would be far worse off: we might stand to lose the immediate benefits of the skill and resource of our engineering and research departments, we might lose easy access to the BBC's training programmes, we would certainly not have the ability to offer staff the kind of movement and experience they can now get in a unified organisation. It is a murky area of doubt and confusion—certainly one that leaves great doubt in my mind.

I am frankly on less secure ground when I speak about the English regions. In television, the BBC's policy towards them has remained based on the idea that the network will happily take their contributions and in so doing ensure that the output is still in touch with its grass roots. But it resists allowing them the right to opt-out from network on a big scale and does not allow them the money to do that anyway. I think the point of real difficulty is that the regions are historical creations and perhaps too cumbersome in their size; North Region with seventeen million people—does the Tynesider see things the same way as the man from Wigan or the Yorkshireman as the man from Cumberland? Moreover, the establishment of the Regional Boards, the establishment of fifteen ITV companies, most of them operating in terms of much smaller areas, have called in question the sheer size of BBC regions. Again, the introduction of the local radio experiment has emphasised the smaller, more intimate audience, potentially at any rate more involved with its station than millions of people lumped together for convenience rather than any clear cognate relationship. At present, the future pattern and financing of radio is under close scrutiny, as is the pattern of television outside London, with particular reference to the English regions. My guess would be that logic would dictate a fragmenting of the BBC regions as we know them into smaller areas for television purposes; and, if the Government agreed to the stabilising of local radio, some development of that on the radio side. If this fragmention did take place, the question of contribution to network might be left open. This seems to me fundamental—

fundamental in the sense that I can understand a regional station, faced with cost factors taking a decision to do its very best with its local programmes, but give up worrying about network contibution; but that decision, of course, means that the mainstream of experiment is left to London or to network companies whose major output, so far has shown little regional basis—were that to happen, it would still leave open the extent to which smaller area units could hope, or be expected, to contribute to the development of the main stream of broadcasting. As I have said, I believe those of us who work outside London must press to join the mainstream; the analogy of the local paper, beavering away at its own parish while the great nationals advance the course of journalism, is not apt to television; there is in the texture of the original *Z Cars* series or *Coronation Street* or even *Dr Finlay's Casebook* in period terms, a reminder to the nation of our own intrinsic quirks and differences, of the rich variety within our constitutional unity. We must seek to be universal against our own environments—as, if I may be forgiven the analogy, were Hardy or Lawrence or, in Scotland, Lewis Grassic Gibbon.

Professor Wedell in his book has floated the idea of BBC 1 becoming a wholly regional channel—admittedly, once a restructuring of the whole broadcasting pattern had taken place. Frankly, I think this is a non-starter. With more than 99 per cent coverage of the UK, BBC 1 must continue as the main national service of entertainment and information—at any rate until such time as UHF coverage of the country matches our present VHF coverage. Nor am I convinced that at any one time, the BBC regions as now constituted would be capable of producing the full fifty-five hours or so a week. It would need a considerable period of regeneration, with frequent blood transfusions of programme monies, before they were geared to that possibility.

On the other hand, regular contributions to the main networks from extra-metropolitan centres (and I am talking now about the BBC) seem to me vital. Vital, because if that ceased to happen, we would, I think, see television turn in upon itself, perhaps rejecting the wayward and the unconventional, striving only for the bland and the easily marketable:

the centralising force of London might then become the only alternative to Hollywood—and its products no less sterile and predictable.

As far as the BBC is concerned, I want to see us strong competitively, bold in experiment and innovation, and courageous enough to accept that many voices, speaking in a variety of accents and dialects, must have access to the camera and the microphone if we can continue to claim to be the *British* Broadcasting Corporation.

8

Management and creativity

Denis Forman

It was a matter of some months between accepting an invitation to speak about the structure of Granada Television and the realisation that what I had promised I had to fulfil. It was at this latter point it occurred to me that I had a problem.

It was this. There was no structure that I could discern. Not unnaturally this was a distressing discovery. Could it, I asked myself, be true? Granada undoubtedly exists, the programmes are transmitted, the shares are healthy, the programme ratings rather good. There must be a structure somewhere.

But although I had lived and worked at the centre of Granada Television for fourteen years, I could not find it. My first desperate thought was that of any worried businessman—call in a management consultant. But further reflection proved that this was an expensive and extravagant thing to do. Perhaps I could myself step outside the company for a few days and adopt at least the demeanour of a management consultant looking at Granada from the outside, and objectively. I did find a structure.

I was surprised and delighted with this new-discovered edifice. It was simple and in some respects rather impressive. I felt as if I had deciphered Linear B. So having found the structure, I will talk about Granada as if I were giving you the notes behind the management consultant's report. They are no more or less than a diagnosis of how the structure of one particular company works in making programmes. This consultant was not asked to recommend improvements, was not asked to condemn or to praise, but so far as possible, simply to investigate and to explain. I had no difficulty as a consultant in acquiring the necessary information, because I had the Joint Managing Director at hand, with his fourteen years' experience with Granada. I make no apology that I confine myself to Granada. That is what I was asked to do and that is the brief I gave to the 'consultant'. It is the only company that he studied, the only company that he knows. There is no necessary relevance to any other company or any other broadcasting group.

The first thing I discovered was that Granada's structure could apply only to a relatively small group of people—which I define as a group quantitatively as large as, say, Granada, ATV, London Weekend, Yorkshire—and not of the size of the BBC, CBS, NBC or some of the Japanese companies. The structure which I found could not operate in any of these giants. It could only operate in a company which was employing in the region of 1,000 to 1,500 people.

The second point that I discovered was that the structure had been evolved by instinct. Instinctive reactions to events had resulted in certain methods of doing things and a certain manner or style of operating. It was clear that these instincts were peculiar to a particular group of people, and were different from those which would have guided another similar but different group of people doing the same job. Therefore the final structure was an 'original'; a 'natural' if you like—different from the structure that any other half dozen people would have devised.

The first major discovery was that Granada is a company operated under a system of dual control. I will illustrate this with a simple diagram.

Viewing Public		Granada Group Board
↑		↑
Programme Committee	....	Television Board
↓		↓
Producers, Directors, Writers	....	Administration

In most television companies matters of fundamental programme policy are resolved at the Board; in most companies those people who make programmes are responsible to the Board. But I found that in Granada there was a Programme Committee, running parallel to the Board, and with a direct responsibility not just to the Board but to the public. This Programme Committee was composed of eight people, all programme executives, who were its constant members. Whereas the Granada Board was responsible upwards to Granada Group, and downwards to the heads of the administrative departments, the Programme Committee was responsible upwards to the public and downwards to the directors, producers and other creative people who make programmes.

Now it might seem a touch irresponsible that the programme-making branch of a company is in no way responsible to its Board. Yet Granada appears to have solved this problem, as it seemed to me, by the simple and rational expedient of having more than half the Board on the Programme Committee, including most of the more senior executive members of the Board. I looked into the records and discovered that no decision of the Programme Committee had ever been questioned by the Board. The reason you will have already grasped. If the Board wanted to query what was happening in the programme area they would then be raising questions about the conduct of half their own members, including the Chairman and the two Joint Managing Directors.

Both bodies have their responsibilities, but the interesting point is that each group conducts its affairs independently. The Programme Committee has, for instance, its own external relationship with the ITA; the Board have a different relationship towards the ITA, and on quite separate matters.

The way this Siamese twin operates is, in fact, simple. The people who sit on both groups, the Programme Committee and the Board, do take on a different attitude and a different demeanour as they walk through the door of one meeting or the other. Their agendas cover two different areas. What they carry with them is the common factor of their own individual personalities.

The next thing I discovered was that there was genuine person-to-person contact on the creative side of Granada. There were about sixty producers and directors and five constant programme executives on the Programme Committee. I realised that, on average, there were twelve producers and directors who were personally responsible to one or other of the Programme Committee. Of course, some executives had more responsibility and some less—and there were several sub-sections running in parallel. For example, a team runs *Coronation Street*, and the programme's Executive Producer is directly responsible to the Programme Committee. But, at the same time, the producers and directors working for him can go to the Programme Committee and then say, 'What about a programme on X or Y?' And since they have all been appointed to their assignment on *Coronation Street* by one of the members of the Programme Committee, they will often want to discuss their progress, their next job, or any other creative matter with a member of the Programme Committee. Thus there is a programme responsibility to an Executive Producer, running parallel to an individual responsibility between an individual member of the Programme Committee and each of the sixty individual producers and directors.

I discovered another point. When major reviews or decisions had to be taken, there was a wide base for discussion in the meetings between the Programme Committee and the creative men and women concerned. If the future of *Coronation Street* were to be discussed, as it has been at regular meetings over the past few years, on average there have been three or four members of the Programme Committee present at a meeting which might last for two hours, along with three or four members of the *Coronation Street* team, the Executive Producer, the producer, sometimes the story editor as well, and one or more directors. When there was a review of drama, or of drama series, I found that all the producers and directors who were directly interested in drama, perhaps fifteen or twenty, would occasionally meet with all the Programme Committee for a considerable length of time. So in addition to the person-to-person direct contact, in addition to the sub-groups, each run by an Executive Producer, there were also wider programme consultations held from time to time.

The next principle was one of fluidity. If you were to take three members of the Programme Committee, and assess their main interests as, let us say, drama and comedy, and the next one's main interest as journalism, and a third member were to have an equal interest in all three, I found that individual producers and directors, according to their own interests, from year to year worked to and with different members of the Committee. A drama director or producer might work for any one of three of the executives on the Committee: equally with a comedy producer. A journalistically-inclined producer might work for almost any one of the Committee.

Of course there were some people who would be likely to stay creatively in the same general area of programmes (because that was their habit and their nature) and with one person. But I discovered that in Granada the concept of moving from department to department or from one professional group to another did not exist; it was more a matter of working with one person or another, or moving from one field of activity to another. In other words, departmentalisation, in the sense of rigid hierarchies, did not exist.

Next I looked at the Programme Committee itself. It met weekly, from about half-past ten in the morning to about half-past three or four o'clock in the afternoon. The meeting included viewing, which might be of some programme that the company had made and which not every member of the Committee had seen. It might be something—a film or TV programme—made elsewhere. I found that their mode of operating was both discursive and executive. That is to say, most areas of interest in public affairs and most new ideas in drama, or in television, or in feature films, would be discussed at one time or another. The end result of the meeting was a number of executive decisions, on average between twenty-five and thirty a week—some of them small and apparently trivial, some of them of fundamental importance to programme policy for the future.

Next I looked at the flow of ideas reaching the Programme Committee. The number of ideas from outside the company was exceedingly small, that is to say from a member of the public not directly concerned with television. But a large number of ideas appeared from people who knew the company, who knew people in the company, or who worked for the company. Very often a freelance working temporarily with the company would talk to a director or producer for whom he was working. This frequently resulted in a programme idea.

The largest number of programme ideas, however, were from the staff. Once these ideas reached the Programme Committee, a large number of them were rejected and the remainder were considerably transformed, except in those rare cases where the first concept had a clarity and a shape so obviously right that it could not, or should not, be amended or adapted. I found, too, that there was a certain body of opinion, both within and without the Programme Committee, which swung from one area of interest to another from time to time, and from year to year, with its point of focus centering variously on political, social or dramatic ideas. This background swing of ideas and opinions tended to generate programmes—often the Programme Committee was the greatest generator of all. By tapping into this psychic radar system and picking up the signals from the people with ideas, it was often possible to produce one final suggestion for a single show, or a series. Sometimes the generation was involuntary, almost a reflex. The mood appeared to be, 'This is the sort of thing we must do' or, 'This is a good idea, we must do it'. And another kind of generation was forced upon the Programme Committee in that sometimes they had an obligation to the network, or to the public, or to the ITA or other companies, which required, perhaps, a half hour programme. Then the generation of course was not 'involuntary'. It was necessary to find a programme: a great deal of time appeared to be consumed on this particular aspect.

Moving from the control of programme content to other forms of control in Granada, I found that in the area of finance the control, although flexible, was very firm, much less fluid and much less informal than the methods adopted in debating and discussing programme ideas. The annual budget was divided under such headings as drama, comedy, current affairs. I also found that within the various headings there was an enormously wide variation between the amount of money applied to one programme and another. Thus *The Caesars* probably needed twice as much money as *The System*, both one-hour dramas which, in most classifications, would fall under the same budget head. Many of the major documentaries had been allocated five times as much money for the same length of time as others. So although the budget headings were firm and laid down, the movement within those heads, in the scope of programmes, was not.

I found that the producer was given a relatively clear idea of the cost-range of his programme at an early stage in his preparation (and I will come to a definition of costs in a moment). Then, having reached a certain point in his planning, he had to report to a meeting where the cost magnitude figure was established for his programme. This was expected to be plus or minus 10 per cent of the final budget—the actual amount of money allocated to a show. This last was produced at a budget meeting just before his programme or series of programmes went into rehearsal. Once the programme was in production detailed costings were returned showing actual expenditure against budgeted expenditure. When necessary, these could be obtained within forty-eight hours.

The weekly budget report indicated programmes which over-spent and those which under-spent. But then I learned from observing the nature of television that the out-of-pocket money spent is not entirely the controlling factor in the cost of a programme to a television company. It appeared that the real criteria of costing and the real governing factor which kept the economy from becoming over-heated was the control of facilities—studios, recording machines, etc. I discovered that Granada had worked out the level of activity which was the optimum for their studio complex and production was designed to fall within this limit (not always successfully). I found that producers, always anxious for the maximum facilities, tended to push the level of demand higher and higher until eventually the more scarce and precious facilities were rationed like bread in a besieged city.

I also discovered a new system of cost accountancy was being introduced which would treat Granada as two companies, one company in the business of transmitting, i.e. purchasing its News from ITN, its films from commercial sources and many of its television programmes from other companies, and its own television programmes from a Granada production company. This second company, the production company, was to be costed as if it were an independent unit. All appropriate charges would be written against it including depreciation, rent, heating, lighting. Its relationship to the other Granada company would simply be that of supplier to customer.

My last point was to examine the criteria that this structure set itself against. What was its aim? What did it think it was making programmes for? And now I would like to drop the conceit of the business consultant and speak in my own persona. After due consideration I put the factors in the following order. The first is a desire to make programmes that match the ambitions and satisfy the creative ideas of the people making them. Most people working in television do not please themselves unless they also please a large public. One massive danger is to give way to the desire to please only the critics and the pundits in the profession. An equal danger is for a producer to make a programme which pleases the public but which he despises himself.

This raises the criterion of ratings and I would like to explain very clearly what I mean by this. So far as I know, the Board of Granada Television has never instructed the Programme Committee of Granada Television to make programmes that will get high ratings. I know that the advertising sales department has never attempted to suggest that programmes should be made simply in order to get high ratings, and I know that the Programme Committee has always, therefore, been free from pressure to make programmes merely for ratings. Now this does not mean that the people who make programmes do not want high ratings. For *Coronation Street* obviously we want the highest rating we can get. The people who make it are proud when they feel they have pleased a maximum audience, and it is their aim to please—and to satisfy—a very big audience. The people who made the Monday play *Ward Six*, an adaptation of a Chekhov story, will be waiting anxiously to know what the ratings are. But they do not wait for any commercial reason. They want to know whether they were good enough to hold the public against Sherlock Holmes. The degree of difference between the numbers who watch Sherlock Holmes and who watch Chekhov is a matter of supreme importance to us, because if nobody watches Chekhov and plays of calibre it might be wrong to go on making them because we are failing to reach our audience.

The best example of this was in the early days. When we said we were going to broadcast *Look Back In Anger*, our colleagues in other companies were not so keen as we were. Some of them, thought we wanted our heads read. However, we put out *Look Back In Anger*—to a rating of 70. In my opinion that was the most important rating that Granada ever got, because it showed that the public were interested in this kind of play. Immediately it opened up a number of new possibilities. Ratings are important from this point of view. They indicate that the public intelligence is often ready to take something new and well produced.

Next, I would like to return to the critics, and in particular to the views of a small number of the best. They are the people whose words have a considerable effect on the makers of programmes. I know myself the effect of a bad notice when it is accurate. It is devastating, whereas the effect of a bad notice when it is foolish, is pure shadow and no substance.

Last I would say that of the various criteria, the least influential is fashionable columnists' talk in the press. You might think at the moment, from reading the papers—which we do continually—that independent television was (a) in an internal panic, (b) had lost most of its viewers, (c) was rapidly putting back all the oldest, corniest programmes it could lay its hands on. This is so far from the truth as one sees it when working in television that it is irrelevant. The same thing used to happen in the war; when you were on a battle front, you would read an account of that battle in newspapers. Usually it was unrecognisable.

So, a final rapid review of Granada's programmes in the last three months. *Coronation Street* seems to have kept its quality and its audience; *World in Action* is having a strong run; we put out six major plays about the *Caesars; Ward Six* (which I mentioned before); and an original by Tom Stoppard called *Neutral Ground.* There were three massive documentaries, *Cities at War*, and a new programme called *Nice Time* that is making some good running. Whatever the popular or the critical views of these programmes may have been (and it has been generally good) they will not, by any terms, construe into the picture of alarm and consternation from which independent companies are alleged to be suffering at this moment.

I hope I have given you a truthful account of the way that Granada works. Perhaps that phrase ought to be 'should work', for we make many mistakes. Our structure sometimes lacks the clarity and the cohesion which the outline I have given you may have implied. We suffer from the usual discontents of creative groups. But there is one thing that I dare affirm: that the vast majority of the people working within the structure have one primary interest—to make the best possible television programmes.

9
Creativity and accountancy

Stuart Hood

Television is concerned with an equation in which the two terms are creativity and accountancy. If that equation were ever to be worked out with complete accuracy it would be a very bad thing. Fortunately it can only ever be worked out in approximations. Accountancy is important because every television station is a factory packed with a complex of extremely expensive capital equipment. It is clearly of considerable importance that it should function economically and that costs should be controlled. Looked at from another point of view, a television station is a complex of people who are—or should be—mostly involved in the creation of programmes. This is not to deny that electronic engineering is creative; it is immensely so. Or, that the best administrators are not, or may not be, creative in their own sphere. But the demands on the creative powers of the programme makers are intense and continuous. Television devours ideas which stem either from individuals or from teams which have worked together over a long period, whose members spark ideas from each other. There is, of course, room for individual thought, individual ideas, but the nature of television is such that teamwork is immensely important, not only in the sense that it is a team—writer, director, designer, camera man, sound engineer, lighting engineer and others—that finally puts the programme on the air; but in the sense that the television organisations consist of creative nuclei from which ideas spring by contact, argument, collaboration. Such nuclei tend to be anarchical in that they wish to create their own rules, to break the regulations laid down by organisers and administrators, to sacrifice procedures to the (to them) more important end of putting good programmes on the air. This is one of the reasons why the equation—accountancy/creativity is so difficult to solve tidily. There will always be pockets of creative chaos which a head of programmes will not, if he is wise, try too hard to reduce to organisational order.

Administrative practice differs greatly between the BBC and the independent companies of ITV. Curiously it seems to have been the BBC, partly because of the immense concentration of studios in the Television Centre in London, that first became aware of the need to introduce the kind of cost accountancy and cost control one associates with the running of an electronic factory. It was no longer possible to tolerate a state of affairs where an idea was presented to management as being cheap, simply because the direct payments involved—the above-the-line costs—were small. A system of costing was therefore evolved which gave each producer an idea of what his programme was going to cost the BBC in terms of manpower, money and resources. Recent recommendations by efficiency experts have merely refined and extended an established practice.

Control both of expenditure and editorial content depends to a considerable extent on communications and line of command. One of the BBC's problems was, and is, that of size. During the development of the television service, departments grew in a somewhat haphazard way until they were very large indeed, to the point where it became almost impossible for a departmental head to oversee all the activities of the producers under him. Some degree of devolution had to come, particularly since the advent of BBC 2 inevitably led to further expansion. Each head of department, for example, was responsible for very large sums of money—was a patron on an immense scale—and it was necessary to know how the patronage was being exercised. The decision was therefore taken to divide these departments into smaller ones which collectively would form a group. Thus there would be a drama group with departments dealing with plays, series, serials. The solution was administratively tidy and had the advantage of forming smaller teams working on specific areas of television. The disadvantages were that the lines of communication became longer; management became more remote.

It is sometimes argued by critics of the BBC that its very size is a negative influence, that the organisational structure of such a very large organisation must be deadening. There is a facile and by no means apt comparison to be made with a Civil Service department. In fact, within the complicated organisation which the BBC has built up during the years there is room for a great deal of initiative and divergence from the norms one might expect to have been strictly laid down. There are some producers who will argue, cynically, that the size of the BBC is the great safeguard for freedom to make the kind of programme one is interested in; that its very size means that not all activity can be controlled; that there are corners in which producers can work quietly at tasks which might be officially frowned on but which may, when completed, be welcomed by the authorities as signs of intiative. To a greater or less degree, most departmental heads are involved in a plot of complicity of this kind. It might be argued that the very size of the BBC, the impossibility of ever bringing everything under strict managerial control, of reaching into all corners, is one of the safeguards of its continuing success as a nursery for talent.

Commercial organisations, like the independent companies, are by comparison with the BBC extremely small. Control is less sophisticated but more brutally effective. Cost accountancy is less highly elaborated than with the BBC but, within a tight organisation, the element of personal intervention by a head of department is more noticeable. Sanctions are also more rigorous. Budgets may be ruthlessly cut. It is a crude form of surgery at times, but it is effective. What is noticeable in the commercial companies is a marked reluctance, in most cases, to give time for thought; time notoriously means money. Most boards of directors are disinclined to invest either money or time in considering future plans, in setting up pilot schemes, in discussing alternatives to tried and successful, but jaded, formulae. On the one hand, therefore, there is closer control and a shorter chain of command; on the other, there is less consideration of what is needed to produce programme ideas—less of that argument and discussion about programmes out of which programme ideas arise from a kind of friction. Since time is so extremely precious and costly, production becomes a pipe-line, a conveyor belt. A producer completes one assignment and has barely time to draw breath before he is plunged into another. Directors fare even worse. One bright director said: 'I have a feeling that I am caught in a long tunnel from which I shall never escape'. These are highly skilled professional men and women who rely on their expertise to carry them through. But being under constant pressure they have no time to stop and wonder if there are not other ways of directing a play or a film, other methods of presentation, other types of programme to be created.

Another characteristic of the conveyor belt system (which shows signs of gaining ground in the BBC) is that directors and producers are allocated to programmes on a rota basis. It is a system calculated to produce programmes without heart. A man is assigned to a programme not because he is interested in the subject or has anything to add to the creation of it, but because he is available and cannot be allowed to be idle and merely think, watch other programmes, study films, or go to the theatre perhaps. It leads to specialists in light entertainment or quiz-programmes being alloted to religious or educational programmes. It leads to cynicism in the directors and to indifferent programmes on the screen.

No one will deny that the management of a television company must have a duty to its shareholders on whose behalf the organisation must be efficiently run. But they have another duty, which is to the viewer; to produce good television. What is saddening, in some sectors of commercial television, is the feeling that so long as the company is reasonably profitable too much attention need not be given to the quality of the programmes provided for the viewer. Long running, established programmes will be retained long after they have lost any spark of vitality or originality because to replace them would be to take a risk, to run the danger of failure. The BBC, on the other hand, has always been willing to take precisely such risks—to experiment and then to back winners. Where cost control has been valuable in BBC terms is precisely that it has allowed money to be deployed for experiment and then for the consolidation of success.

Accountancy can be used in either of two ways. As a tool to make creativity possible or as an instrument for the sale of commodities. Marx defines a commodity as something which satisfies human needs, irrespective of whether they spring from man's belly or his imagination. A commodity has value not because of what it is but because of its exchange value. There is a growing danger in British television that programmes will come to be thought of as commodities, articles for sale—preferably in the United States. Some companies are dominated by commodity merchants. In their view a variety show is a commodity; so too is Ibsen, provided it is introduced by one theatrical knight and stars another; so too is an indifferent telefilm—provided always that it finds a buyer. Their judgments are neither aesthetic nor social but strictly commercial. As British television faces the present financial crisis there will be increasing pressure not only in the independent companies, but in the BBC to produce commodities for the world market. The result can be that the British viewer loses his importance.

In terms of creativity the danger of the commodity approach is that it calls for safe formulae, for repeatable formulae. Experts are called in who believe that a programme can be taken apart like a clock, stripped and reassembled. They are immensely efficient, very skilled; but their product is somehow dead. It lacks the heart-beat of the work that has been produced by the obsessive creative urge of an individual or a team, by that excitement and desire to communicate to a large number of one's fellow-countrymen, which lies behind the making of programmes in any truly free television system. Both the commodity merchants and the programme makers are interested in audience size—in the ratings. The first because they know almost no other method of measuring the success of a programme; the others, because it is important for them to know how successful they are in communicating thoughts and emotions to the audience. Ratings are immensely important to the programme maker—as proof of contact, a confirmation that he has not worked in vain. A high official of the BBC once said that there are some programmes worth broadcasting even if no one is listening. That seems to me a kind of blasphemy.

10

Between consumer and producer

B C Sendall

I quote from the Pilkington Report of 27 June 1962:

Unless and until there is unmistakable proof to the contrary, the presumption must be that television is and will be a main factor in influencing the values and moral standards of our society.

Giving oral evidence, Mr. Carleton-Greene, the Director-General of the BBC, considered that television would be one of the main factors influencing the values and moral attitudes of our society. We need only add that nothing in the Corporation's written and oral evidence has suggested that they hold any reservations about the principle.

Not so the Independent Television Authority. I quote:

The Authority was disposed to regard as exaggerated the commonly held view that television exerted a high degree of influence. That it would be a main factor in shaping the values and moral attitudes of our society in the next decades was, in Sir Ivone Kirkpatrick's* submission, an over-simplification. He thought that society would be largely what it was—with or without television. He would hesitate to accept the dictum that society was going to be in any way moulded by it. . . . In fact the Authority told us that, though there was a duty to widen tastes and affirm standards, the broadcasters' task was essentially to mirror society and its tastes as they are.

One more quotation, this time from an interview given by Lord Hill, as Chairman of the BBC, to *The Sunday Telegraph*, 1st September 1968:

Sir Hugh Greene has done a fine job in liberating the BBC from its old image. That needed to be done and that he has done. . . . He has achieved a transition to a service that seeks to mirror life as it is—warts and all. That needed doing if broadcasting was to become sensitive to the rapidly changing mood of the people.

* At that time Chairman of the ITA.

There was a curious happening within weeks of the evidence given by Sir Ivone which so disturbed the Pilkington Committee. The Home Secretary, then Mr R. A. Butler, concerned about the growth of delinquency, called a conference of representatives of the Churches, educational and social service interests, the press, the broadcasting services and others, to see what might be done about the moral climate of society. In one respect—perhaps in one respect only—that conference was not entirely sterile. This was the offer by Sir Ivone to provide from ITA funds, the money for research which would attempt to establish some facts about the impact of television on society. It took the Home Secretary quite some time to get to grips with this sudden proposition. There was a conference of social scientists and broadcasters at Sunningdale in the Spring of 1962 which recommended the formation of a committee to initiate and co-ordinate such research. That committee—the Television Research Committee—first met rather more than a year later under the chairmanship of Mr Noble, the Vice-Chancellor of Leicester University, in the knowledge that it would have at its disposal a quarter of a million pounds from the ITA over a period of five years. I would like to come back to this committee and its work later. For the moment it is enough to note that, after six years, nobody still seems to be able to say with any solid evidence to support him whether television is to be seen as a major factor in forming the moral standards and values of society, or is indeed more to be seen as mirroring society as it is, as thought Sir Ivone in 1961 and as thinks Lord Hill at the present time.

Any reference to feedback conjures up visions of committees, and of course, the broadcasting bodies are surrounded with advisory councils, committees and panels. The Pilkington Committee was surprised that the ITA had not, like the BBC, appointed a General Advisory Council and recommended that it should be required to do so. In the event, it was not required to do so, but it did do so; such a council has been in existence for five years. The Committee did not say very much about the use of research techniques for ascertaining public opinion about the television services, but it did recommend that the ITA should engage—as the BBC already did—in what it called 'continuing and perceptive audience research', the cost of which should be an admissible charge on revenue. This recommendation was followed up in the Televison Act of 1964, which included the making of arrangements for ascertaining the state of public opinion concerning the programmes as one of the statutory functions of the Authority.

However, the Pilkington Committee itself had made no use of research techniques in its own attempt to assess and appraise the public mood about broadcasting. It seems to have thought it sufficient to invite evidence from whoever chose to submit it, and justifiable to regard what it received as a true reflection of the public mood as it then was. What this evidence largely lacked, and what the evidence that any comparable committee appointed in future might well also largely lack unless it departs drastically from traditional procedures, is objective and definitive information either about the social impact of broadcasting or about the attitude and response of the community as a whole, to the programme services with which it is provided.

In his chapter on 'The Consumers' in *Broadcasting and Public Policy*, Professor Wedell has ably traced the development of the institutions which are designed to guide the broadcasters regarding the needs and desires of the audience they serve. In addition to Parliament itself and the part-time laymen who make up the governing bodies, he refers to a 'whole range of additional activities, official and other . . . whose purpose is to reflect the wishes of the consumer in relation to broadcasting or to act as a link between the broadcasters and one or other of their constituencies'. He identifies four groups: the official advisory bodies, general and specialised, central and regional; the unofficial groups, of which the best known today, of course, is the National Viewers' and Listeners' Association (VALA) the audience research activities of the broadcasting organisations; and lastly the press critics who, he says, 'may or may not regard their task as including the provision of a link between the producer and consumer'. Professor Wedell himself argues his way through—a little hesitantly perhaps—to the concept of a single national broadcasting advisory council which, he says, 'should be so constituted and related to regional councils that consumers at large may feel themselves to be adequately represented'. Rather a tall order this, I feel, however skilfully the appointments are made. Because he is concerned lest his own ideas about devolution in broadcasting might be thought to run counter to this concept, he returns to it at the very end of his book and adds the suggestion that the machinery of audience research might also be operated by an independent body closely related to the national advisory council.

There is, of course, a fifth group that Professor Wedell might have identified. It is the largest, if not necessarily the most influential, of them all. I refer to that vast club which has no committee, no chairman and whose members always pay their subscription in postage stamps—that very disunited association of viewer correspondents. What sort of people belong to it? Well, we recently analysed at the ITA the correspondence received from viewers over a couple of months. Three letters out of four came from London and the Home Counties where about one in four of the television audience lives. Three out of five letters were from women—and 90 per cent of these women were married. Judging from such indicators as vocabulary, style, use of headed notepaper and so on, the vast majority of writers were middle or upper-middle class, whereas, of course, only 30 per cent of the general audience are middle class.

This, of course, highlights a characteristic of all but one of the sources of consumer advice that have been mentioned. In the last resort the advisory bodies, official and unofficial, the viewer correspondents and the critics represent nobody but themselves. They belong to the legions of minorities that may, or may not, on some occasions merge into majorities. Our experience over five years at the ITA of working with a General Advisory Council has been a wholly beneficial one—we have had some marvellous discussions with it of a kind which we could not have had in any other private forum available to us—but, as its members would probably be the first to agree, we could in theory have ten or even a hundred different General Advisory Councils and have equally valuable, though appreciably different, discussions with each. And when Mrs Whitehouse of VALA writes to Lord Aylestone Chairman of the ITA, as she sometimes does, to say that she has had a lot of complaints from members of her Association about a programme which she may not herself have seen, she cannot really know for certain whether she is writing on behalf of her Association or not, bound together as its membership may be by a common devotion to certain defined standards of behaviour and taste.

Audience research, however, operates on a very different plane, for it claims to distil by means of scientific sampling the opinions and attitudes of all the consumers everywhere. It tries to tell us not just what some people think and how some people react but *how many* of the whole audience think one thing and how many think another. Of course, the more refined the evaluations it attempts to elicit, the more cautious it is about generalising from the responses of the sample; but all audience research breaks down unless a high degree of trust in the representative nature of the sample can be maintained.

Now I will frankly confess that I am sold on audience research. I would say that it is much the best technique for securing viewer participation that we have got or can ever hope to have. I was sold on it even before I read Dr Belson's formidable work *The Impact of Television* which I would describe as a beautifully structured, hard-selling commercial for extending the use of research techniques throughout the television industry. Like every enthusiastic salesman Dr Belson sometimes oversells. He is apt to assume that audience research alone produces facts. 'Decision making by those in television' he says in his opening paragraph 'will be good and successful only if it is based on the relevant facts'. In his final paragraph, 369 pages later, he is still saying it: 'Only when broadcaster decisions and public pressures are guided by *facts* is there a reasonable chance that the constructive potential of the television medium will be approached.' In between these two punch lines there is a lot of good solid argument to show what benefits would stem from the greater use of research in all its aspects. I would like to dwell for a while on some of these aspects as they present themselves to me.

As one of those shadowy ex-civil servants who haunt the darker recesses of ITA and sometimes incur the wrath of a Wilfred Greatorex or Peter Black, I have spent a good deal of my career poised precariously between the consumer and the creator. Thirteen years have now been spent in television, but before that, I played not dissimilar roles in other fields. I am sold on audience research because I know it can, and frequently does, provide information which is unobtainable by other means, and which, once one has got it, one wonders how one could ever have done without it. From our own experience within ITA I might quote two examples in the area of educational television. Our adult educational advisory body has for sometime wished to move from a purely sanctioning role towards a more positive policy-forming role. Its members were, however, conscious that their experience, extensive though it was, related to institutional learning, and the students with whom they were familiar differed substantially in many respects from the audience for adult educational programmes on television. We therefore carried out a survey using rather special questioning techniques, on a representative sample of 2,000 adults throughout the country, in order to establish the degrees of interest which were expressed in an exhaustive list of 130 possible topics for educational broadcasts. As a consequence of this survey we now know what groups of topics, and what individual topics, are of greatest and least appeal: to what sections of the potential audience they appeal most or least: at what level of difficulty each should be treated: regional variations in interest patterns, and a wide variety of other detailed data. We are in a much better position to formulate a policy for a planned adult educational output, designed to provide a balanced schedule catering for both majority and minority tastes. Only research could have put us there.

Secondly, in the area of schools broadcasting we have never really known what use was made of our considerable effort. The liaison with schools which has been established by the programme companies has provided information of a qualitative kind which, while useful to those responsible for the production of programmes, has never provided sufficiently *systematic* information about the reactions of either teachers or pupils, or about the *amount* of viewing which takes place. We therefore undertook a feasibility study, in which several different approaches were tried out, in order to see which would yield the most reliable and useful results. As a consequence of this study we now know how to obtain accurate and reliable information on how many schools, of what kinds, in what areas, are using our schools broadcasts: of what the opinions of teachers are about each series and each programme in a series: and of the attitudes of pupils towards the broadcasts, as well as the information learned through them. It remains for us to decide how intensive our research effort should be, having regard to the costs involved. The point I am making, however, is that information such as this, essential for the appraisal of our efforts in this area of programming, was not obtainable through the existing links for feedback, and can only come from systematic research.

As well as providing information which is not obtainable in any other way these two research projects possess the inestimable advantage of being *useable:* and we would not have undertaken the one, and be planning to undertake the other, if we did not know that the people running our educational television were ready and willing to make use of them. The television industry is already, I suspect, littered with unused, and in some cases unuseable, research reports. I believe that the risk of non-use greatly increases, the further away from the sources of programme production the motivation for doing the research is to be found. Three years or so ago we had a great success with some rather elaborate research into the tastes and preferences of the audience for television drama. It has had a marked influence on the development of ITV drama policy since then—an area in which we have remained strong during a time when in some other areas our strength seems to have been dwindling. I am certain that the origin of this success is to be found in the care with which in this particular case we consulted company production heads before embarking on the project.

That is why I think Professor Wedell is wrong in wanting to see audience research in the hands of an independent body linked to an independent national advisory council. Dr Belson is nearer the mark when he tries to pinpoint the reasons for the continuing lack of rapport between researchers and producers which is impeding the wider deployment of research techniques in television. Social researchers are, or should be, the cooks who turn the basic ingredients of feedback into a digestable diet. I would say, for that matter, although it is not within my main theme, that Professor Wedell is also wrong in wanting a single, independent, national advisory council. The people who make television are consumers too: they are the actual or potential consumers of advice. And advice is a commodity which has to be skilfully marketed if it is to catch on. The nearer to the point of sale the advice can be brought, the greater the chance of the advice being effective.

By the test of useability the audience measurement service based on meters that used to be supplied by Television Audience Measurement Ltd. and is now supplied by Audits of Great Britain Ltd. is of indisputable value, as doubtless also is the service conducted by the BBC employing the technique of aided recall. The programme planners are provided with a continuous record of the number of homes which have viewed their programmes and also the programmes transmitted by BBC 1 and BBC 2. They also know whether a programme which starts off with an audience of a given size maintains that audience throughout, or ends up with an audience higher or lower than the one with which it started. It is sometimes thought that audience ratings are crude measurements. Actually, they are the very reverse of crude, and the planners are on sure ground with them so long as they always keep clearly in their minds what it is that the ratings are measuring and what it is that they are not. They do not measure popularity: they measure the success or otherwise of a programme—having regard to the time at which it is shown—in securing an audience against the competing attractions of the other programmes on the other channels. The programmes in the upper reaches of the Top Ten may well be the most popular programmes, but they are by no means necessarily so, and there may well be others much lower down the list which are intrinsically more attractive to viewers. This feedback of the measurements of size and composition of audiences, whether they are made in terms of sets switched on or in terms of people who viewed, is an indispensable tool of the programme planners.

Whether, by the test of useability, a parallel audience *reaction* service is of comparable value may be more open to argument.

The BBC has undertaken research of this kind for many years, whilst Independent Television has largely relied upon regular public opinion polls and special research projects designed to help in the solution of *ad hoc* programme problems. ITV now means to move into line with the BBC and equip itself with an audience reaction service which will produce reports furnishing a simple index of enjoyment or appreciation of all programmes viewed by members of a representative panel, together with much more sophisticated and detailed evaluations of specially selected programmes. Our aim will be to base our entire evaluation procedure upon the terminology and vocabulary spontaneously used by the viewers themselves, rather than upon the classifications and descriptive jargon which are conventionally used by the broadcasters. We are hoping over a period of time to provide ourselves with an additional tool based upon the ways in which members of a representative panel of viewers evaluate programmes and react towards them. We shall have to guard against it being too unwieldy for the programme planners or the programme makers to use to advantage. Broadcasters are only human, and they react as all humans do to the pressures and frustrations of their work. If the precision tools of the trade are too fussy and finicky, and place too great a strain on their patience, then they will put them aside and resort to chancier methods based upon personal hunch and their own native do-it-yourself dexterity. I am not sure that researchers are sufficiently mindful of this. Just as programmes sometimes get made with little or no regard for the audience for which they are intended, so also research is sometimes done as an end in itself, without enough thought to its capacity for being put to effective use.

In conclusion I would like to end where I began with a reference to the social impact of television, reporting upon which must surely be a fundamental element in feedback. I drew attention to the classic Pilkington obsession about television as a *major influence*, if not *the* major influence on our society, and contrasted this with Lord Hill's reference to Sir Hugh Greene as liberating the BBC from its old image (the one which Pilkington so warmly approved) into a service that seeks to mirror life as it is. And I described how, in an odd sort of way, the Television Research Committee came into being on the rebound from the Pilkington–Kirkpatrick clash over this very issue. Was that issue a real one or was it not? Whereabouts in the social spectrum does television really stand? We in television would indeed be happy if we could be told. We would not like to go on for ever being required to be all things to all men. Can research come up with some sort of answer? Or is the verdict to be, like the verdict of history, that there *is* no verdict? Mr Noble and his team will, I understand, soon be making a report on their five-year-old assignment.* It is eagerly awaited.

* Since published, see Bibliography.

11

The critic's role

Philip Purser

To clear up straight away one or two things a television critic *is not*, and therefore what this chapter is not going to be about: first, surely it is a diminution of the function of any critic—in any field—to regard him as merely a tipster or a consumer's guide. So I will not waste time chasing the old hare that because a television critic mostly writes after the event—after the programme is done and gone—he is of no practical value to anyone. How many of those who read a theatre critic have access to a theatre, let alone the London theatre he is discussing nine-tenths of the time? How many filmgoers can recall what Penelope Mortimer, or Penelope Houston, or Penelope Gilliat said about it by the time it gets along to the Odeon, Nerdley?

All critics exercise influence, of course, but they exert it on small sensitive areas rather than on the whole public: theatre and film critics on exhibitors, ticket libraries and the inner core of London cognoscenti who can tide a show over in the West End until a wider reputation builds up; the book critic (alas), I was going to say, on the librarians and fashionable booksellers, but really book criticism is something apart now: it is more a substitute for reading the book than a comment on it. The TV critic's pressure area is made up of producers, programme controllers, governors, even chairmen of companies. One of the most valuable instruments of criticism ever applied was Peter Black's habit, in the early days of commercial television, of resting the responsibility for the disgraces on the remote figures who were in fact responsible—the Prince Littlers and Spencer Wills.*

Nor are all television programmes 'over and done with' in a night. They are alive as long as they reverberate on in someone's memory. Many are physically repeated, many more—the

* The original chairman of ATV and the chairman of Rediffusion Television respectively.

majority, indeed—are not single items anyway; they're part of some continuing series. A series can often be seen to respond to intelligent criticism—the way *The Plane Makers* evolved for instance, or *24 Hours* in its early days, shedding the resident panel of pundits which was such a bore. Barry Took has told me that *Bootsie and Snudge* resulted from some observations of mine about these characters in *The Army Game*. Good notices can get a locally-seen show extended to the whole network—I am credited or blamed, whichever you like, with getting *Nice Time* a wider audience. Performers and writers have thanked me at times for something I have written which, they say, has led to them being considered for some other, prized assignment.

But what is most emphatically not the critic's job is to supply the craft criticism which producers, directors and—in particular—designers and lighting engineers and so on, sometimes demand. It is no concern of mine if camera three should have been swung to give a two-shot in the teashop scene instead of the crane shot that was used. I am not going to detail off a section of my mind to keep a routine watch on make-up and scenery and costume. If it is a heavily atmospheric or period or horrific piece in which these things are playing an obvious positive part, okay; otherwise a useful rule is that if you notice them they are probably wrong. When a novel of mine is published I do not need to be told of a misprint on page 158: the publisher's editor should have spotted that. What I want to know is whether the whole thing has worked or not, whether the ideas have got through. If BBC 2 put on *Billy Budd* it is nice for them to be told that the descending sevenths or whatever they are in bar 92 were as immaculately played as ever they have been, but what they really want to know is if they succeeded in overcoming the enormous basic obstacle to televised opera—the difficulty of persuading what is essentially a matter-of-fact medium to absorb such an outrageous convention as two grown men singing secrets to each other at the tops of their voices.

If we have to use the word 'feedback', and I am not very happy with it, then this kind of feedback is obviously of value. I once likened the plight of the TV producer to that of a practical joker who had just dyed the whole of London's or Manchester's water supply bright pink, and was waiting for the reaction. The water board could tell him how many taps had been turned on during the joke, and the number of customers who took the trouble to ring up. Only some of them thought the colour was blue, one thought it was Pepsi Cola and three were convinced it was part of the fluoride conspiracy. Only the critic could tell the producer whether it was a nice pink or a nasty one and whether it was a better joke than the man the year before who had put guppies in the water.

I am still going to suggest that the televison critic's chief contribution is not to the producer, nor even to the audience, but to the *act* of television. It sounds rather pompous and patronising but the critic completes the act. He is a sort of midwife—or madam. Someone puts out something on television. Lots of people see the something, the critic legalises the exchange. I do not mean that people are not perfectly capable of watching all by themselves and making up their own minds about what they have seen without the benefit of highbrow—or, for that matter, lowbrow—advice. Of course they are, or lots of them anyway. And for a few people living quite alone the relationship with television must be as simple and innocent and undefiled as that. In the normal course, though, watching television is not a solitary experience. It is not a group experience either, of course. One is not a member of a crowd sharing crowd emotions. It is a molecular experience: two or three watching together, or at least talking together about the programme the next day at work or on the bus.

Views about the ideas and images, the merits and demerits of performers—in the widest sense of the word—are being constantly exchanged. There is nothing I resent more about Mrs Whitehouse and all her works than the assumption—which flies right in the face of her campaign for greater participation—that the television audience is an inert field of cabbages being crop-dusted with poisonous influences. There is a continual positive reaction among the recipients—you only have to listen in a pub or a supermarket to hear it. We critics are simply recipients who publish their reactions, who are paid to justify them rationally, to think them out further, test them against a rather more conscious backlog of experience, to argue them more closely, perhaps, to express them more cogently. We do not—or at least I do not—profess for one moment to be spokesmen. I do not speak for anyone but myself, which is a very important point to which I shall come back. I do hope that some people will find they agree with what I say, just as I know some will violently disagree. I hope that mostly I am in the van of whatever progress TV is making. I might even hope that I am attracting one or two others in that direction. I would have no objection to being called a guide or interpreter.

Let me amplify this a little. Like any other work of art a television programme should be capable of giving some satisfaction and pleasure without calling on any prior knowledge on the part of the beholder. But no one would suggest that the pleasure and satisfaction cannot be doubled or even trebled if some of the depths of meaning below the surface are pointed out. To give an obvious instance, the play which echoes some old legend or story: the other week there was a spy play by Tom Stoppard based on the story of Philoctetes; this was declared: but I remember a Kingsley Amis re-working of the *Duchess of Malfi* which wasn't declared, or many years ago now, a play by Ian Dallas which was a covert version of the story of David and Bathsheba. In neither case was the source obvious and only one or two critics spotted it. In neither case did it matter very much whether you did spot it. But I think that if I had not known I would have been interested to have this key, and it would have lent colour to whatever musings I had about the play afterwards. I think that similarly I would appreciate being told that a striking piece of film in the *Ten Days that Shook the World* or *Cities at War* came in fact from a feature movie and not from a newsreel, or that *Coronation Street* was doing a sly parody of Jean Luc Godard, or that the roots of a Michael Bentine comedy sketch could be traced back to a gag he first experimented with in 1961.

Promote this task from one of simple annotation to the zone in which a piece of television's basic credentials are under consideration and you begin to arrive at the critic's chief usefulness. Television no longer—if it ever did—simply observes the world; increasingly it manipulates the world or is manipulated by it.

Increasingly it ventures into the twilight areas between fact and fiction, between reality and illusion, between event and non-event. It is a minefield, really, through which you have to pick your way pretty carefully. The critic is the man who's paid to go ahead and put down the markers.

I remember a marvellous moment on ITN a few years ago when in the middle of the bulletin they went across live to Switzerland, and under the Alps, for the moment when the two teams of tunnellers driving the St Bernard road tunnel linked up. You saw someone breaking through the last thin rock wall, and then all the handshaking and toasting. It was television at its purest—the old ideal of see it happen as it happens. The cameras were there to observe what was going on anyway, they did not affect what was going on.

Or did they? Was it not possible that the engineers were ready to go earlier? Is it not human nature to want to make a little ceremony of a moment of achievement? We have all seen the same thing happen on election nights as camera-conscious returning officers await their cue.

Another tunnel: there was an exciting NBC documentary about a tunnel under the Berlin wall, to which the climax was the genuine escape of genuine refugees running an authentic risk of capture and even death: a very exciting piece of *cinéma vérité*, and containing little moments of truth that no feature film writer would have thought of, such as the first reaction of some bedraggled, muddied young mothers on reaching safety—which was to change the nappies on their babies. But there were also certain indications, certain gas-leaks of the sort the critic needs to have a nose for, that this was not quite the piece of privileged eavesdropping it purported to be; and of course it transpired that the escape, however desperate for those taking part, was set and financed by the broadcasting company, and indeed that all the American networks were in fierce competition to see who could organise the most thrilling one.

Even the smell, or should we say the tone of voice, of a compilation programme assembled at the rostrum and the editola needs to be questioned sometimes. I am no historian but I fell out violently with the compilers of *The Lost Peace*, which has lately been repeated on BBC 1, for it seemed to me that the years following 1918 were invested with despondency that could not possibly have been as universal and relentless as this—it was even raining in every shot!

A different kind of example: a young Japanese newsreel cameraman thought he would make a film record of his father's first experience of hospital, where he was expecting to have an operation for a minor internal obstruction. It was discovered that the old man had cancer and was dying. The cameraman decided to go ahead with the film anyway, on the grounds that others should know. The result was shown on Japanese TV and at a number of TV festivals. I found it moving and reverent and also, of course, harrowing. Undoubtedly the young Japanese was persevering in a task which he believed to be honourable. But again the conscientious critic needed to ask himself—and perhaps thus voice doubts which some of his readers had sensed but couldn't quite isolate—had the technical challenge of the film led him on just as much as the idealistic motive? A director was listed in the final credits in addition to the cameraman himself. What part had he in this filial reportage? What whispered consultations had been necessary around the dying man's bed?

Something else again: last month we had the third showing of *Cathy Come Home*, everyone's favourite Dear Me drama—by watching it and saying 'dear me' we all feel better. Well of course it is an affecting story, with a very moving performance by Carol White. It would take a hard-hearted observer not to be distressed by her final rock-bottom despair as the children are borne off into care. It would also require a very docile observer, I believe, not to feel he was being got at by the author and producer. Among the credits this time were a number of streamer slogans. One said reproachfully that West Germany had built twice as many houses as Britain since the war, which did not seem very relevant and may be ignored. Another said that 4,000 children had to be separated from their parents every year because of situations similar to the one which the play ends on. A third pointed out that every incident shown had actually occurred somewhere in the previous eighteen months. Shocking facts, but surely an admission of failure on the part of the play if it has to protest it is based on fact. The truth we look for in fiction—even documentary fiction—is not the kind of truth to to be verified by reference to the Registrar General's annual report. The truth we look for is the kind which tells us how people behave and how their behaviour affects what happens to them. This is just what *Cathy Come Home* did not offer. Cathy and her husband were not people at all, they were victims—with a capital 'V'—subjects to whom things happened, things moreover that were part of a relentless conspiracy: every landlord hostile and adverse to children; every ratepayer—when their caravan encampment was under pressure from the locals—a lynch-happy fascist; every welfare worker unsympathetic and bound by red tape.

This conspiratorial view of society is one which I happen to distrust, not so much from political conviction as from plain scepticism. I do not believe man's inhumanity to man can be expressed in simple black and white terms—how easy it would be to solve the problems of the world if it were so. In anyone's experience, the most complex mixtures of prejudice and stupidity and kindness and confusion operate. But is it not also part of a general and unwelcome dehumanising process? Sweeping statements, manifestoes, audience research, sociology are all so much easier to organise if you can categorise people as—consumers or viewers or ratepayers or whatever it is. It is much easier to present any situation that turns up as a clash between two mutually exclusive camps—between management and labour, between blacks and whites, or socialists and conservatives.

I reject all attempts to present people in this way. Indeed I reject and despise all audience classification and audience measurement. I have no time for sociological surveys, which either strive massively for the obvious, or confirm that the lowest common denominator of humanity is indeed low and common. I regard with abhorrence the proposal of Dr Belson that television production should be governed by audience research. We have the example of literature programmed to readership research. Its apostle is Mr Bernard Geis. Its most celebrated product, I suppose, is *Valley of the Dolls*.

Television unfortunately is very susceptible to this depersonalisation, with its reliance on audience figures and its weakness for bogus sociology. It plays the categorisation game itself, dividing people into believers or non-believers, into motorists, or non-motorists, or—as in Cathy—into victims and victimisers? There was even one series, *Sanctuary*, which divided us into nuns and non-nuns. I believe that the television critic's most important job of all is to promote the absolutely selfish, solipsist, individual view—both in his response to TV programmes and in his attitude to television as a medium and an industry and all the other things.

Television is something which on the whole is addressed by very, very few people—perhaps a dozen or two dozen at a time—to even fewer people: two or three on average. They are not an audience, they are not a crowd. The individual, personal, feedback is the only one that matters. The person who provides it is conventionally known as a critic.

12

Patterns of broadcasting structure

E G Wedell

A discussion of patterns of broadcasting structure must derive its orientation from the more detailed concerns which were discussed in the preceding chapters.

Nonetheless, this is a reflex mechanism which causes those concerned with broadcasting in one way or another to use the existing time-honoured structures as their point of departure in a discussion of broadcasting policy. In this chapter I restrict myself to an attempt to identify some of the considerations that should go into the making of broadcasting structures, and to the sketching in of no more than a broad outline of some criteria which derive from the earlier chapters.

Broadcasting as a community concern

When in the autumn of 1968 there was the short-lived rumpus about broadcasting and the politicians, Mr Wedgwood Benn was much berated for saying that broadcasting is too important to be left to the broadcasters. I cannot understand why this should have been so. It seems to me to be one of the more self-evident truisms about broadcasting, as about other matters of public policy. Surely none of us, in the last resort, would be content to see defence left exclusively to the generals, the determination of transport policy to the railways, conduct of the Press to the newspaper proprietors; and so on. Of course we leave to them the day-to-day running of these services, as long, and only as long, as we are satisfied with their performance. Let them go wrong, and we begin to assert in no uncertain terms the right of the tax or rate payer, the right of the transport user, or of the reader, to express a view.

And this is as it should be. The fact that, in broadcasting, the tradition of participation which was developed in the age of sound radio has not carried on into the television age does not mean that what the broadcasters do, or leave undone, should not be subject to the scrutiny of people other than broadcasters.

One of the difficulties in broadcasting, as elsewhere, is that there is little *steady* critical concern (in the neutral sense) by the non-professionals with broadcasting. (I here include the governors and advisers among the professionals, although I am aware that they are appointed as lay representatives of the public stake in their various organisations.) On the whole criticism, of whatever kind, becomes vocal only at points of crisis. A new departure in programming may be one cause; an inflammatory speech about the broadcasters another; the sitting of a broadcasting committee a third. On such occasions hotheads tend to come out on top, emotions run high and judgments are formed in the bowels rather than the head. In present conditions, where news coverage of an event cannot usually be sustained for more than three days at most, the fuss ceases as suddenly as it began, usually at the very stage at which people may have thought about the issue sufficiently to have some useful views to contribute.

The Benn–Crossman pother was a good example of such a short-run crisis. I am not anxious to disinter this, particularly since the arguments seem to me to have been conducted on too narrow a front. Moreover, as Dr Blumler and Dr McQuail in their *Television in Politics*, have recently shown once again, politics on television does not necessarily deserve all the attention lavished on it.

What the row did indicate was the vulnerability of broadcasting to such partial and short-term judgments; and the need for two ingredients to any pattern of broadcasting structure: firstly a more sustained method of keeping broadcasting issues under review; and secondly the presence of sound and resilient structures of broadcasting which can withstand short-term fluctuations of public, or indeed partisan, interest and taste.

The need for thought about structures

But, it may be said, why worry? After all, the existing broadcasting organisations are reasonably thick-skinned. They can withstand quite a lot of bullying. And, as far as can be discerned, there is no great upsurge of demand from the public for participation in broadcasting decisions. Nonetheless I believe that there are good reasons for giving careful thought to the evolution of broadcasting structures.

I have mentioned the harm, actual and potential, done by the off-the-cuff judgments arrived at during short-term crises. There also remains, in spite of much negative correlation deduced from research into the mass media, the nagging worry which runs right through the Pilkington Report and which formed the background to that Committee's recommendations. As they said, 'the background to our thinking was the realisation that broadcasting will have profound and far-reaching effects; the working assumption must be that television in particular will be a potent factor in influencing the values and moral standards of our society'. I suppose nothing that has happened in the last six or seven years has really reduced that worry, although the context and the nature of it may have changed.

Again, the financial problems force themselves on our attention. As a former Postmaster-General Mr Short, said in 1967 'the dichotomy between the public and the private sectors creates a financial problem of such dimensions that it is difficult to see the present arrangements continuing for more than the next decade.' And Mr Short promised himself a long cool look at broadcasting as soon as he should be freed from the day-to-day administration of the Post Office when the new Corporation was established.* In fact the current arrangements, the Television Act 1964 and the BBC Charter, expire in 1976. Taking into account the time taken by necessary legislation, and before that by a study of the questions that arise and the necessary argument about them, the time scale does not leave much leeway.

So we have to face the fact that, whatever the broadcasters may wish, they will not be left alone for long to get on with the job. Hence a discussion about long term objectives may not be out of place at this time when the pressure for decisions is not quite so immediate. If we could rely on another genius like Lord Reith to deal with the problems for us, that would be comforting. As the effective architect of the present system and of its ethos (whether in the BBC or in ITV) he created a

*In the event, Mr. Short was moved shortly afterwards and the job now devolves upon the new Minister of Posts and Telecommunications.

coherent structure, whatever we may think about its fitness for our present, or our future, purposes. And he made the system according to his lights, largely because no-one else bothered to stop him. As he has said, 'Odd sort of muddle-up it was thought to be; neither one thing nor the other; not worth bothering about; wait to see what a mess there would be. . . . And the thing worked; against most all expectations it worked; I saw to that'. But conditions have changed and it is, I suppose, a good deal more difficult for any one man to get away with such a stroke of genius; certainly not in an estate of the realm such as broadcasting has become.

So we can but make corporate attempts to discern the signs of the times and to pattern our structures of broadcasting accordingly. What do the signs for the 1970's portend?

Firstly, *the technical possibilities have opened out*, not as much as some would wish, but nonetheless sufficiently to allow a good deal more diversity of broadcasting than has been the case in the past. Even if not all the technological wonders mentioned in chapter 1 become practically viable in the foreseeable future, we have room, assuming the continuance of a policy of national or near-national coverage for all public services, for at least four open-circuit vision services and four or five open-circuit sound services. By the end of 1969 the allocation of the fourth channel should again be under discussion. The UHF vision and VHF sound coverage areas are small enough to enable us, if we so wish, to differentiate programme output between quite restricted geographical areas. The future of wired services is more uncertain, but we are told that the standard cable will carry six vision and three sound channels: larger cables are a possibility. The practical contributions to broadcasting in the UK of transmission by satellite seem to be a little further off.

The technical possibilities are matched by *a swing in the socio-political pendulum from centralism to devolution*. It is quite clear that we cannot regard the tendency for concentration into larger and larger units in some economic and administrative spheres as being necessarily axiomatic in the social or cultural spheres. We are only just beginning to adapt our thinking to the new but, I suppose, basically simple fact that remote control in these spheres is ineffective, and inappropriate to the scale of human living.

Thirdly, *the consensus of spiritual and moral orthodoxies* by which our public life has been informed until fairly recently now seems to be in progress of final dissolution. It is interesting to trace the break-up of this consensus from the time in 1943 when it was still possible to envisage the remarkable agreed provisions concerning religious education embodied in the 1944 Education Act; through the period when the BBC, in giving evidence to the Beveridge Committee in 1949 was able to speak of broadcasting as 'a bastion against the tide seeking to submerge values in a disintegrating world'; and via the BBC's Pilkington evidence, in which no comparable passage occurs; to the present day, when it is regarded as natural that an appeal letter in *The Times* about Biafra should be signed not only by the Archbishop of Canterbury and other ecclesiastical dignitaries on behalf of the Christian churches but also by Professor Ayer as President of the British Humanist Association.

Alongside these more general socio-cultural changes has gone the cautious but discernible development of a more experimental approach to some of our accepted broadcasting norms. The attitude to broadcasting hours, which used to be compounded of a mixture of 'the more the better' by ITV, 'more, with more money' by the BBC, and a holier-than-thou attitude by governments unwilling to raise the licence fee; appears to be softening. The prim assertions of the December 1966 White Paper are unlikely to be repeated, that

> 'In the Government's view, the amount of broadcasting time will remain a matter of sufficient social importance to require that the PMG should continue to hold and exercise his powers of control. Nor do the Government consider that any general increase in broadcasting hours will be justified for the present. They do not, however, rule out the possibility of more time for education programmes.'

And in the next paragraph the Government gave the go-ahead for Radio 1 and to broadcasting until 1.30 a.m.!

Similarly, as Sir Sydney Caine suggests in chapter 3, it is now becoming possible to explore, though as yet with a good deal of hesitation and under firm public control, sources of broadcasting finance other than the licence fee, for the general run of broadcasting services.

Against the background of these changes and this loosening of positions which were thought only a few years ago to be immobile how will our present patterns fare in the years to come?

A choice of structures

The answer will depend on that to another question: what sort of broadcasting do we want? Different people have given different replies in the past few years. Let us look at two or three of these.

The Pilkington Committee wanted a minimum of changes in the case of the BBC; and the closest possible approximation to the BBC pattern of the Independent Television system. This recommendation did not commend itself to the Government of the day. It is just worthwhile after six years to pause and to ask whether, from our present stance, this is what we would after all want to see.

Although he presumably endorsed the recommendations of the Pilkington Committee, Professor Richard Hoggart in August 1965 postulated, in *The Listener* a scheme of what he called *democratic* broadcasting, and distinguished from *state controlled* and *commercially impelled* broadcasting. Hoggart recognised that democratic broadcasting 'is, by its nature harder to operate and is not really loved by establishments and big battalions, political and commercial, since it has an awkward independence of them. The essence of democratic broadcasting is the effort objectively to serve as wide a range as possible of diverse interests.' Neither of the broadcasting organisations in this country is wholly state-controlled or wholly commercially impelled. So the gain to broadcasting from a move in the democractic direction is likely to be less the relief of ills arising from these impulses, than an access of such qualities as awkward independence and the capacity to serve a wide range of interests. It would be useful in the near future to pursue these principles and to explore what their precise structural implications might be.

Again, in an interesting article in *The Listener* in July 1968, Mr Raymond Williams tried his hand at a proposal for reform of the structure. He complained that 'any orthodox system maintains itself by limiting the terms of available choice, by confusing the language in which what it is really doing can be described and, under pressure, by raising devils we don't know to enhance the devils we have. The ping-pong of public corporation (BBC, monopoly) and independent corporation (advertising, private capital) has gone on until most of us are mesmerised. . . . but the break to freedom that is necessary means getting beyond the established assumptions'.

Mr Williams then proposed producer-oriented broadcasting organisations coupled with a scheme of elective authorities which 'while similar in function to the existing authorities, would be essentially different in that they would have no powers apart from their component elements: the producers and the public would be directly in charge without the intervention of an appointed and self-defining authority. The means of finance would be wholly public, through licence revenue and direct grant'.

It is easy for broadcasting professionals who know all the snags to dismiss proposals of this kind, made by students of the social and cultural scene who are not at the same time skilled administrators. Their schemes are often, as they stand, impractical and visionary. But it would be dangerous to dismiss the ideas which are put forward in these ways. They contain ingredients of the live issues which we discussed in the preceding chapters of this book: the relationship of the creative man to what is inevitably to some extent bureaucratic, namely a broadcasting organisation; and the relationship in the same context between producers and their audiences.

So I would go with Hoggart in seeking to make broadcasting more responsive to the needs and interests of its constituencies, and with Williams in seeking a better deal for the creative people, even though I do not think either of their schemes as they stand are workable.

What else do we want from our broadcasting system? Do we want more broadcasting? Do we want more choice? Do we, within that choice, want better quality of programmes, or more relevant programmes, or more programmes tailored to regional and local interests? Do we want more of what we call freedom for the broadcaster? Do we want less, or less influential, broadcasting? One could expand indefinitely the list of possible desiderata.

However complete the list, and wise our choices, the trouble is that ideas change. Even if our own were to remain constant, those of our successors will differ and they are unlikely to seek our advice. In addition to identifying our wants as clearly as we may there will, therefore, be virtue in paying attention to the rendering of our structures sufficiently flexible to allow them to adapt to needs which will change faster in the future than they have done in the past.

Which patterns are likely to meet this specification? I want to throw out for discussion three criteria, rather than the patterns themselves, which carry a good deal of weight with me and which I believe should guide our pattern making.

The first brings me back to the beginning of this chapter: the importance of setting broadcasting policy firmly and permanently into the context of social and cultural policy (in so far as this is now emerging, c.f. the policy for public subsidy to the arts), so that we do not find ourselves undoing under the terms of one act of policy the good that we do through another.

Secondly, the development of structures which enlarge the circle of those able to participate responsibly in broadcasting is equally important. There is no need to labour the compensatory role of broadcasting institutions at a time when the number of outlets for the expression of opinion in the press is contracting. Unless a community can talk to itself easily and with vigour, democracy cannot survive.

Lastly, the enlargement of the area of freedom for the broadcasters is of equal importance, particularly in the multilateral cultural era into which we are moving. I would myself be prepared to consider some sacrifices in other directions in order to make this possible. Devolution of control in broadcasting so that the number of independent sources of output is multiplied would, for instance, seem to me worthwhile, even though this might reduce the 'national' character of some broadcasting services. The reduction of some broadcasting to a more human scale would be valuable, even if it were to make for more fallible broadcasting. Standards of excellence are important, but we must ask ourselves whether they are in all cases of supreme importance. It is at least conceivable that a broadcasting system might be less technically and professionally perfect than our present one, and yet socially more creative and more calculated to promote the common good.

Bibliography

Bibliographies of material about broadcasting are rarely satisfactory, since the subject covers such a wide range of interests. The books listed below deal with some of the themes discussed in the preceding chapters. Those marked * contain bibliographies which will help the reader who wishes to go more deeply into the subject.

Report of the Committee on Broadcasting, 1960, Cmnd. 1753, (The Pilkington Report) HMSO 1962

Coase, R. H., *British Broadcasting: A study in Monopoly*, Longmans 1960

Hood, S., *A Survey of Television*, Heinemann 1967

*Scupham, J., *Broadcasting and the Community*, Watts 1967

Simon of Wythenshawe, E., Lord, *The BBC from Within*, Gollancz 1953

Jenkins, C., *Power Behind the Screen*, MacGibbon & Kee 1961

Wilson, H. H., *Pressure Group*, Secker & Warburg 1961

Hoggart, R., *Mass Communications in Britain*, in Ford, B. (ed.), *Pelican Guide to English Literature No. 7*, Penguin 1961

Williams, R., *Communications (rev. edition)*, Chatto & Windus 1966

Halloran, J. D., *The Effects of Television*, Leicester UP 1964

Klapper, J. T., *The Effects of Mass Communication*, Glencoe, Free Press 1961

Swallow, N., *Factual Television*, Focal Press 1966

Windlesham, Lord, *Communication and Political Power*, Jonathan Cape 1966

Trenaman, J., *Communication and Comprehension*, Longmans 1967

*Briggs, A., *History of Broadcasting in the UK*, OUP, Vol. 1, 1961, Vol. 2, 1965

Caine, Sir Sydney, *Paying for TV*, London, Institute of Economic Affairs 1968

*Wedell, E. G., *Broadcasting and Public Policy*, Michael Joseph 1968

Halmos, P. (ed.,) *Sociology of the Mass Media Communicators*, Keele, Monographs of the Sociological Review 1968

*Blumler, J. G. & McQuail, D. *Television in Politics*, Faber & Faber 1968

The Yearbooks of the BBC and the ITA (annually) are also excellent sources of reference material.

Index